REMINISCENCE
of a
ROVING SCHOLAR

Science, Humanities and Joseph Needham

REMINISCENCE
of a
ROVING SCHOLAR

Science, Humanities and Joseph Needham

Ho Peng Yoke

Director Emeritus, Needham Research Institute, Cambridge

World Scientific

NEW JERSEY • LONDON • SINGAPORE • BEIJING • SHANGHAI • HONG KONG • TAIPEI • CHENNAI

Published by

World Scientific Publishing Co. Pte. Ltd.

5 Toh Tuck Link, Singapore 596224

USA office: 27 Warren Street, Suite 401 402, Hackensack, NJ 07601

UK office: 57 Shelton Street, Covent Garden, London WC2H 9HE

British Library Cataloguing-in-Publication Data
A catalogue record for this book is available from the British Library.

ISBN-13 978-981-256-588-4
ISBN-10 981-256-588-4

Printed in Singapore

To

my wife Lucy

Contents

Foreword

We see the past as the future will see us.
- the *Lantingxu* 兰亭序 -

The above quotation comes from the *Foreword to the Orchid Pavilion* (*Lantingxu*), written by Wang Xizhi 王羲之(321–379), the most celebrated calligrapher in Chinese history. It subtly warns against making hasty judgement before one understands the historical, social and cultural background of the subject in question. A case in point are Chinese emigrants to Southeast Asian countries during the first few decades of the last century when they could hardly come to terms with the changing political and economic conditions in their homeland and found themselves in a new political, social, cultural and economic environment.

In 2001 I visited Ipoh after an absence of 40 years. Ipoh is the capital of the State of Perak in Malaysia and was the place where I received my primary and secondary education. I went there to give a talk at the Perak Academy. There I met a friend, whose sister was married to a professor in Cambridge and I suggested that the Academy might try to get that Cambridge don as a guest speaker some time in the future. My friend's father was a wealthy tin-miner in the Kinta Valley that once produced two-thirds of the world's output of tin. Among my hosts was a barrister, who later told me a story about him. The barrister said that the tin-miner was once his client in a civil lawsuit and found himself confronted by cultural conflict. His client took the attitude of an employer, saying that since he would pay for the service, his legal adviser should say in court what he wanted him to say. After listening to the case, my informant advised his client that the latter's presence in court was unnecessary and that he could handle the case all by himself. Not knowing that the advice for him to stay away from the courtroom was for his protection, the tin-miner turned up in court. He quickly lost the case after being put to answer a few questions by the opposition's lawyer. This story is not meant to be a mockery of the misadventure of my friend's father, but rather to illustrate a great handicap arising from cultural differences that Chinese immigrants inadvertently had to concede to some other members of the communities.

I am in full sympathy with my friend's father, as his case finds parallel with that of my own father. My father rented two adjoining double-storied townhouses in Ipoh in 1934 and used the top two floors to run a traditional style private Chinese school. The school stopped operating when the Japanese army occupied the town in December 1941. He rented out one of the townhouses to a sub-tenant. When the war ended, he wished to reopen his school and made every effort to reclaim what he had rented out. He thought that he had a very strong case with the support of both the education department and the representative of the owner of the two properties. His lawyer presented his case with all these points, but was ruled out by a post-war byelaw of the local Town Board that safeguarded tenants and sub-tenants from being ejected by the landlord or chief tenant. Had he known about what was in the law, my father would have saved much time and money over the matter.

When "common sense" (*qingli* 情理) based on Chinese culture conflicts with local law that contains a strong element of Western culture, the latter inevitably prevails. Without knowledge of the official language of the country he lived in, my father was unable to put into practice the ancient Chinese motto, *ru jing yi wen jin, ru xiang yi wen su* 入境以问禁, 入乡以问俗 — to learn about the customs and the law of the land wherever one goes to — that he used to teach his pupils. The above two cases exemplify one of the handicaps conceded by the Chinese non-labour immigrants in a multicultural society. As a child, I was brought up in a Chinese migrant family that understood only traditional Confucian values, yet was educated in an

English language de La Salle Christian Brothers' school that taught me values which were sometimes rather different from those I had learnt from my father. After leaving school I found myself in a multi-cultural society, being exposed to cultures and values dissimilar to what I was previously taught. I learned that accommodation was the name of the game for survival in that environment. My ambition after leaving school was entering university to read for a degree, although I realised that it would probably turn out to be wishful thinking, because my father had no means to support me. Neither did I have any specific subject or university in mind. Engineering, science, and the humanities would do — I dared not think about medicine, because it would take too long for my parents to wait for my support. As for the university, I hoped to get into one in Hong Kong, if not Singapore, failing which one in China would also do — getting a university degree in a British or American university was far beyond my dreams. I began my tertiary education and then my academic career during a period of social and political change in the most southernly part of the Asian continent. I received training in physics and became a Lecturer in Physics in the University of Malaya in Singapore, but was soon called upon to collaborate with Dr Joseph Needham, FRS (1900–1995) of Gonville and Caius College, Cambridge in the *Science and Civilisation in China* project. Three years after a promotion to the Readership in the Physics Department in Singapore, I received an offer to take up the Chair in Chinese Studies at the University of Malaya in Kuala Lumpur. Forty years later, my former university in Singapore, where I worked in its Physics Department, asked me to give a talk on how I managed to turn myself into a Professor of Chinese from a physicist and on what a physicist could do in Chinese studies that a Sinologist was not expected to be capable of doing. I talked about the importance of adaptability and illustrated what a physicist could do better than an ordinary Sinologist with several examples of research in Chinese studies that involved astronomical and even astrological calculations and interpretations.

Since 1953 I had been working with Needham, sometimes closely and sometimes over a great distance for over half a century. We were from entirely different cultural and family background and yet I managed to adapt myself to work with him without being too compliant. My early training in

Chinese was a critical complement to Needham's talent in certain areas of his *Science and Civilisation in China* series, while my early social contacts had played a role in the fundraising exercise for the Needham Research Institute. My early upbringing in both traditional Chinese and Western cultures enabled me to work with Needham.[1] Since I had known Needham over a long period and had observed him through my own understanding of both cultures, my observations may therefore differ in certain aspects from other Western and Chinese writers. It is not a matter of whichever is correct, but rather that views taken from different angles may be refreshing to the reader.

A number of friends have assisted me in bringing this book into fruition. The book itself came about at the suggestion of Dr Phua Kok Khoo 潘国驹 of Singapore. I have sought the advice of Dr Tai Yu-lin 台玉玲 and Mr Lim Ho Hup 林和合, both of Singapore but originated from Malaysia, to verify facts relevant to those regions. Mr Liew Yin Soon刘衍存 and Mrs Christiana Liew 高艳霜, both of Brisbane and formerly from Malaysia, helped me to read through every sentence in the book. So did my daughters Sook Kee 淑姬 and Sook Pin 淑苹. Lastly, in chronological order, Professor Sir Geoffrey Lloyd and Dr Christopher Cullen of the Needham Research Institute, Cambridge, had read through the whole book and offered their invaluable suggestions. To all of them I am indebted.

My academic career made it necessary for me to travel around the four continents of Asia, Europe, North America and Australia. Frequent travels that often spanned long period of time entailed corresponding absence from home. I could not have given my wife and our children the attention they so much deserved, but they had not only put up with my omission ungrudgingly but also given me help and encouragement in many ways. This book is written in the year of the Golden Wedding Anniversary between Lucy and me. To Lucy 冯美瑶 it is affectionately dedicated.

---

[1] I once overheard a remark made by the Head of a certain College in Cambridge saying that I must have been a very easy person to get along with, since I could get along with Needham. That Head of College was a friend of mine. I said, in a light vein to a mutual friend, that the remark came too late for me, as that Head of College would make an excellent referee, but at my age I could no longer have the opportunity to take advantage of it.

1

Early Years and Parentage

My knowledge about my parentage is extremely scanty. My father said that his great-great-grandfather, He Jieping 何芥屏 went to Panyu 番禺 district in modern Guangzhou city from Shangyu 上虞 in Shaoxing 绍兴 district in Zhejiang province. In those days, many scholars working as advisors for local magistrates in South China came from that region. My earliest known forebears were perhaps among their ranks. Henceforth my ancestors regarded themselves as people from Panyu district. He Jieping's son, He Guanshan 何冠山 gave him four grandsons, namely He Wenyuan 何文元, He Wenguang 何文光, He Wenkuan 何文宽 and He Wenyao 何文耀. He Wenguang was my great-grandfather. He Junqing 何俊卿 was my grandfather. He had two surviving sons and a daughter. Being in the service of the British custom office and working along the coast of the Pearl River, I would guess that he could speak some English. My father He Qihan 何其汉 (later known as Ho Tih Ann[2] 何迪庵 and he also used the literary name

---

[2] Since this was the English version of the name he adopted personally and used officially, it is not rendered here under any other romanisation system. The same applies to all other similar cases. Most of the personal names for Chinese in Malaysia and Singapore are their official names in English. Sometimes they have no Chinese equivalent, as in the cases of my friends Ivy Soh and Beda Lim. Needham once tried to give Ivy a Chinese name and Beda had also asked me for one. However, the later additions had never come into official usage, such as in passports and banking accounts. I do not know the Chinese equivalents of the official names of some of my Chinese friends in Malaysia and Singapore, including some of those mentioned in this book.

Hancha 汉槎) was born in China in 1895. He had an elder sister He Qiaozhen 何翘珍 and a younger brother He Qijie 何其杰 who served in the local police force. When he was young, my father went to school to prepare himself for the civil examinations. He excelled among his classmates in essay-writing and used to write two different essays for one topic set by the teacher, so that he could sell one of the pieces to a classmate to earn some pocket money. To him the abolition of the civil examination system in China was a disappointment. He had great admiration for Zeng Guofan 曾国藩 (1811–1872) and Liang Qichao 梁启超 (1873–1929), and collected their writings. He recommended me to read *Jinsilu* 近思录 when I became older. Thus I deduced that his thinking came under the influence of the late Qing scholars and neo-Confucianism.

My father also had a cousin named He Qixun 何其埙, who owned a flourishing law firm in Guangzhou until 1949. I had contact with Aunt Qiaozhen and Uncle Qixun together with members of their families, but did not have any news about Uncle Qijie.[3] My father married my mother Wu Peiyao 吴佩瑶 when she was sixteen. My grandfather passed away when my father was only a teenager. It seemed that in those days, the post in the custom office could be passed on from father to son, but it went to a relative instead because my father was not yet of age then. My father then joined the army as a civil officer under Chen Jiongming 陈炯明, a powerful warlord in South China during the first quarter of the 20th century, and rose to the rank equivalent to a Major General. His direct commanding officer was probably General Ye Ju 叶举. During the third decade of the 20th century, military factions in South China were often at odds with one another. Chen Jiongming did not come to the support of Sun Yat-sen 孙逸仙, although, according to my father, he was not totally against the latter. In 1925 my father left the army for the Malay Peninsula. He said that he felt sick seeing Chinese soldiers killing their own kind in battle. He changed his name to Ho Tih Ann and no one in his land of adoption had ever heard of his original

---

[3] In the English language, the word "uncle" when applied to Uncle Qixun here is merely honorific and not a strict description of kin relationship. In Chinese custom, the honorific "uncle" is normally used. To indicate kinship, when occasion demands, the prefix *tang* 堂 is attached.

personal and literary names again. My mother came to be known under the name Ng Yeen Kwai 吴贤贵.

I had an elder brother whom I had never seen, because he died in China as an infant before I was born. I was the second son. My school records showed that I was born in 1926 in Papan, a tiny little town in the Kinta Valley in Malaysia that had seen its days of glory as an important tin mining area, but is now almost forgotten. A younger sister came after me. Unfortunately she died as a child in Ipoh Hospital after a fall resulting from a sudden lightning strike while she was playing outside the *attap* (thatched) house in which my father had rented a room. The accident occurred around 1928. That was also the year when my younger brother Yew Lam 耀南 was born. In 1930 I had a

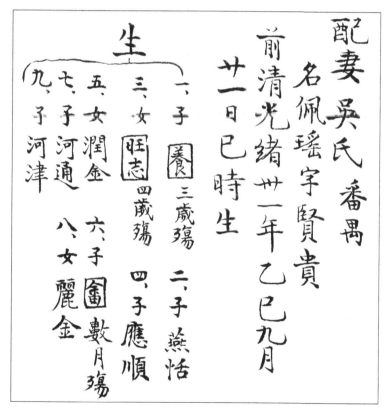

Figure 1 Names of my mother and her children in my father's handwriting. Note that my name Peng Yoke is unlisted.

new sister Ah Looi 亚女. Thus I had a younger brother and a younger sister in my childhood days; we were all too young to be aware of the considerable hardship that our parents were going through. My mother later had five more children, but only three of them survived. She lost her fifth child when he was only a baby and her last birth was a stillborn. Hence our brother Ho Thong 河通, sister Lay Kum 丽金 and brother Ho Chuen 河津 joined the family in 1934, 1936 and 1938 respectively.

Chinese names often cause confusion to Westerners resulting from cultural differences.[4] It was customary, especially among the *literati*, to have a number of names besides their personal names, such as literary names, appellations, titular names, etc. Also commonly used were childhood names and even nicknames. Take my own case for example. I received my literary name Peng Yoke 丙郁 when I first went to school in 1933 and henceforth it has become my official name. Previously my parents called me by the name Yantian 燕恬, or more often Yatian 亚恬 for short. I also had a more formal personal name Yuhua 煜华 that shares a common character *hua* 华 with the formal personal names of all my brothers. However, none of us had any occasion to use those names.[5] Similarly the formal personal names of my sisters share the common character *jin* 金, read in Cantonese as *kum*, meaning gold and also daughter in this context. My second sister's name is Runjin 润

---

[4] Later on, in my academic career Needham wrote my name as Ho Ping-yü, using his own modified Wade-Giles system of romanisation. My publications in Chinese in the Mainland carry the *pinyin* form He Bingyu. My first publication in Britain bears the name P Y Ho, but Ho Peng Yoke appears in the majority of cases. Change of order of family and personal names is unavoidable living in places where Western culture pervades. In Australia my name is often written as Peng Yoke Ho. That is a correct form, but I am sometimes called Peng Ho, without knowing that, unlike Western names, Peng and Yoke must come together and are not independent names. Some other Asian names also give rise to confusion. For example, I found the name of a friend, Tan Sri Raja Haji Aznan, listed after his father's name, Raja Haji Ahmad, in the *International Who's Who* (1989 edition), taking the father's name as the surname or family name. The bibliography had little choice, as Muslims do not have family names. The same applies to Indian names.

[5] Although all these names should be read in Cantonese, they are put in the *pinyin* form, because no English version of these names was in use.

金, as she was born in an intercalary month, but Ah Looi that simply means "daughter" became her official name, because it is the name shown in her birth certificate. My parents had not yet thought of a name for her at the time of registering her birth and supplied that name only provisionally. Getting a birth certificate was something unfamiliar to the traditional Chinese. I cannot blame my parents for not fully understanding the importance of such a document as the birth certificate. My brother Yew Lam was born a few days earlier than what is entered in his birth certificate — because of public holidays the report was made several days later — so that his birth certificate indicates the date of reporting instead. I do not know what exactly was written on my own birth certificate, which I was told, was lost during the war. Chinese put family names before personal names, contrary to Western practice. A personal name often consists of two characters. In Hong Kong and Southeast Asia it is rendered in English as two words, while in the Wade-Giles system the two are connected by a hyphen and in the *pinyin* 拼音 system they appear as a single word. Rendering a Chinese name in two separate words often mislead Westerners to take the two words as two different names.

Despite problems arising out of cultural difference, Chinese names sometimes have more to tell than their Western counterparts. First take the case of my own name, which is rather uncommon. One day Professor Ho Pingti 何炳棣 and I had a personal chat and discovered that our ancestors came from the same district in Zhejiang province, China. As we had often been asked whether we were related to each other, he wondered why the first Chinese character in my personal name was unlike his, without the usual *Fire* radical. My answer was that since I lived in the south, which represented *Fire* and the Chinese character concerned also represented *Fire*, to add a *Fire* radical to it would bring about excess of *Fire*. He nodded in appreciation and said that my father was knowledgeable about the art of fate-calculation. My father often talked about physiognomy, the art of reading fortune from facial observation, but I do not know how much he knew about the Chinese art of fate-calculation or how much he believed in it. At least I can tell that he had some fancy for it. He had a detailed horoscope cast for himself by an expert in Hong Kong and a simple one done for me by another in Ipoh around the middle of the 1930s. His horoscope contained about ten pages and said that he would not reach his 67th year and that ten years would be added to his life by good deeds on his part. He passed away in 1957 at his 63rd year. By Chinese custom three years can be added to the age of a deceased person, so that he can be regarded to have reached his 66th year. The accuracy of his horoscope depends on its interpretation. My horoscope was written only in a few lines, ending with the remark that I would be inclined towards literary pursuits and would have asset of the worth of three hundred thousand. There seems to be something in the first part of the remark, but the last part is so vague that it can mean anything. Nevertheless, these two horoscopes were my first encounters with the method of fate-calculation, which became a subject of my research 50 years later.

The names of my brothers and sisters also reveal somethings about my father. My brother Yew Lam had the name Ying Shoon 应顺 or Yashoon 亚顺 before he went to school. The character *shun*, meaning smooth, indicates that my mother had an easy birth, but also bears the hope that life would turn smoother following the birth of my brother. My sister's name, Runjin, had another meaning besides indicating that she was born in an intercalary month. By adding the *Water* radical to the word *run* for the intercalary month, it also has the meaning of remuneration or money that my father was in want of in those days. My father was hoping that fortune would change with the arrival of his daughter. The name Ho Thong, meaning an unobstructed river, showed that the way was cleared and that life had already begun running smoothly for my father. When his next daughter came he was in the mood to give her more affection and named her Lay Kum, meaning beautiful daughter as well as beautiful gold. The name of his youngest son Ho Chuen has the meaning of a river ferry, indicating that the child was born during the better years of his life.

My father earned his living by teaching traditional Chinese. Sometimes he had very few pupils to teach at home and at other times he had to give outside private tuition. He had to struggle several years before he could settle down in one place to earn a living. I vaguely remember my childhood days when I followed my parents moving from place to place within the Kinta Valley in Perak that was once the biggest tin producing area in the world. Aunt Qiaozhen was married to Chin Wing Hing 钱永兴, a former *juren* 举人 degree holder who had once served in the army. He left China for Lahat, a small town in the Kinta Valley, to set up a private school bearing his own name. Perhaps that was the reason why my father made the Kinta Valley his choice. However, both of them was not very successful in their occupations. My aunt and her family moved from Lahat to Falim, a small town outside Ipoh with only one row of houses facing the main road, and finally to Ipoh itself. As for my father and us, I remembered moving around Pasir Puteh (a village of Ipoh), Tanjong Rambutan, Lahat and Falim, before finally settling down in Ipoh town itself. I was too young to remember the proper sequence of these moves. My father once went back to Hong Kong briefly, leaving my mother and me together with my younger sister behind in Tanjong Rambutan, probably to attend the funeral of Chen Jiongming and to explore any possibility of finding another career in China. However, I was not aware about his actual plans. Anyway he did come back to us.

My earliest recollection of dates was New Year's Eve of 1929. We were at my aunt's home in Falim when my cousins said that we were in the 18th year of the Chinese Republic and that the next day would be the 19th year of the Republic. I also remember that my father gave me two cents asking me to buy one stamp to post a letter at a nearby post office. The post office gave me two one-cent stamps. Thinking that only one stamp was required, since my father asked me to buy one, I posted the letter with one stamp on it and returned the other to my father. My father did not get angry; he said that the recipient would pay for the postage due.

Around 1930 my father took us to Ipoh. Between then and 1932 we moved from one place to another, living as sub-tenants in rented rooms. It was in one of the rooms we moved in that I first saw electric light. Before then I knew of only kerosene lamp and candles. Around 1932 my father moved to a room in a clan association and held classes within its premises. One day, I accompanied him to the Confucian association to attend a talk given by a distinguished scholar, Chen Huanzhang 陈焕章, a Master Degree holder from Columbia University. Of course I could not understand what the talk was about. I remembered only the names of the speaker and the university, and was impressed by the university degree the speaker possessed.

My School Years

In 1933 my father rented a two-storied townhouse at 52, Yew Tet Shin Street, Ipoh and established the Sze Mu 时务 Private School, using the top floor as classroom. The name of his school suggested the influence of the late Qing thinker Liang Qichao. He employed Wan Choy Wan, a school leaver from St. Michael's Institution, to teach English. It was then that I first learned some English. My father decided to send me to an English school. In September 1933 Wan Choy Wan took me to St. Michael's Institution and enrolled me temporarily in a Primary Two class.

By 1934 my father expanded his school by renting two adjoining two-storied townhouses at numbers 28 and 30, Theatre Street, which were within a stone's throw of the original campus. His fifth son arrived to mark this occasion. The two upper floors served as classrooms, one ground floor became our living quarters and the other ground floor doubled as a playroom for table tennis. He soon employed two domestic servants to help my mother. My youngest sister and my youngest brother arrived during these happier days of his life. Apart for a very brief period at 52, Yew Tet Shin Street, I was brought up within these premises during my school years.

My father conducted both day and evening classes, using only one classroom for the latter. He employed two teachers to help in the day school, one to teach English and one to look after one of the two classrooms. In 1938 Lee Weng Toon 李荣端, a distant relative (a cousin by marriage of my Uncle Qixun) and a trained primary school teacher in China, joined the school. Being single, she stayed with my parents for many years until she left to teach in the Sun Yat-sen School in Ipoh. The evening class took in mainly English school students and adults, who were otherwise occupied during daytime. I sat in during the evening class, sometimes helping those who found difficulties in their lessons by explaining to them in English. In this way I had contact with students from English schools other than those from the school I attended.

I owed the knowledge of classical Chinese entirely to my father. He acquired a complete series of library books for elementary schools, entitled *Xiaoxuesheng wenku* 小学生文库, published by the Commercial Press,

Shanghai and put the few hundred books under my charge. I read almost the whole collection and had probably gained the knowledge expected from one who had gone through a Chinese primary school. Some of what I had learned later turned out to be applicable in my academic career. My father also had a copy of the Chinese translation of J.A. Thomson's work, with the Chinese title *Hanyi kexue dagang* 汉译科学大纲 (Outline of Science in Chinese Translation) (Commercial Press, Shanghai, 1923). That was my early introduction to science. The following two examples show what I had learned.

In 1935 my father had a very eminent guest, Zhu Ruzhen 朱汝珍, the *bangyan* 榜眼 title recipient in the last civil examination to be held in the presence of the Manchu Qing emperor. I offered him tea in my father's study, addressing him as Venerable Teacher (*laoshi* 老师) and stood by the side of my father to listen to their conversation. The guest wrote a scroll — to commemorate the occasion — bearing the four characters, *jiao si wu qiong* 教思无穷 (teaching and thinking without limit). My father had the scroll framed and displayed on the wall. One day a friend, on seeing the scroll, made a remark on the excellence of its calligraphy. My father concurred with his remark, but added that a writer with the stature of a *bangyan* could have developed his own style of calligraphy, rather than keeping the style adopted officially for civil examination that was copied from some famous calligraphers. The lesson I learned was that I could learn from others by all means, but I should develop my own style and not merely become a copycat. Recently, I saw in a satellite television programme about a public interview conducted in China with the new CEO of General Electric. To the question whether he would perform in the same way as his very successful predecessor, he replied that he would certainly refer to his respected predecessor, but would work in his own personal style. He reminded me of the remarks my father made on calligraphy.

My father used to tell his pupils his secret of writing. He could write multiple versions on a single essay topic when he was in school. It was simply to put things in proper sequence. He gave the example of building a hut. One had to erect the pillars before working on the roof. This sounded rather obvious at first, but many years later it rang in my ears while I observed Joseph Needham at work.[6] His remarks on calligraphy again resonated in my ears when I later studied the methodology employed by Needham to write his monumental work *Science and Civilisation in China.*

My father experienced both generation and cultural gaps. He had the reputation of being a good teacher, but he was also a teacher much feared by his pupils. Nonetheless many parents sent their children to him to instill more discipline in them. He also applied the principle of sparing the rod to his own children, but only to his first three surviving ones. However, his attitude changed altogether in my case after I attended English school and a close father-and-son relationship developed. My parents regarded their children as their own financial assets. They tried to plan their children's career, including marriages, although not all of their efforts were always successful. Any pressure on me came through my mother, but never directly from my father. As the eldest surviving son, I was supposed to carry the heaviest responsibility, by looking after my parents as well as my younger brothers and sisters when needed. My sister Ah Looi attended my class when I taught at the Sze Mu Private School in 1941. I once taught my brother Ho Thong in 1945 at St. Michael's Institution and at the University of Malaya ten years later. I had my sister Lay Kum in my Physics class in 1957 at the University of Malaya. In 1955 I took my youngest brother Ho Chuen and our cousin Chin Wai Hin 钱维显 to St. Michael's Institution for enrolment as new pupils into the primary class. Hence during my younger days there was some kind of a dual relationship between my brothers and sisters and me. Only my brother Yew Lam was there to play with me. I remember the days when we went together for *kung fu* sessions in the morning at the Ching Woo 精武 Association and in the afternoons when we played table tennis at home. I do have the affection of my sisters and the other brothers, but I feel the loss of not having any opportunity to play with them, especially in the case of my youngest brother whom I know was quite skilful in table tennis. For the responsibility I carried I have earned the respect of my brothers and sisters. My brother Yew Lam also carried much responsibility on his shoulder. He

---

[6] See under the Methodology of Joseph Needham.

lived with my parents for a much longer period than I did. I left home in 1946, while he was with them throughout their lives. However, this is not the place to tell his story.

My father seemed to realise his cultural handicap. Perhaps this was the reason why he sent his children to English schools. The choice of school for me was made not by him but by an old boy of that school. For my brother Yew Lam he picked Anderson School, because he heard that it was a government school and those from such schools would find it easier to enter government service. He did not send my sister Ah Looi to school, because he felt that a woman had no need to find a job and would rather have her stay at home to give my mother a helping hand. The choice of St. Michael's Institution for my two other brothers was entirely my decision, when I was then teaching in that school. I could not remember the reason why my father sent my youngest sister to the Methodist Girls' School. Perhaps it was for variety, and my mother had no need for more helping hands at that stage. During the last years of her life, my mother told me that the best thing my father ever did to me was to send me to an English school.

After only a couple of months in Primary Two class, I obtained almost full marks in all the subjects. I came out top of the class in the final examination and gained a triple promotion to Standard Three class. There was only one incident that I remembered in the Primary Two class. I was introduced to the football game for the first time. The teacher divided the class of about 40 into two sides to play against each other. Being ignorant of the rules, I stopped the ball with my hands to help my goalkeeper. That resulted in my side losing the game. When I got to Standard Three class I found my name skipped over in selection for games and sports. Perhaps my mishandling of the ball was not forgotten or perhaps I was one year younger than average in class and had a non-athletic look. But to me it was a blessing in disguise, as I had sufficient exercise from my early morning *kung fu* session and my late afternoon table tennis game with my brother Yew Lam. Besides, not having to go back to school for games and sports enabled me to participate in the evening class session in my father's school. Moreover, I even found time to take up fretwork as a hobby. I often worked my file through the plywood while humming the tune *'Tis a long way to Tipperary'* that I heard and learned from the school scouts. That this was a popular British military song was unknown to me and so was the location of Tipperary, but it did serve to make me forget the monotony of working the saw up and down and keep me on my way.

I became myopic in 1934, but took some time to realise the need to correct my vision with spectacles. Being only exposed to spoken English for less than half a year as compared to my classmates, who had already been in school for four years, my deficiency in that area soon showed up. My classmates jokingly called me "Chinaman" and qualified it with an adjective *chinkjong* that one might interpret as the derogative word "chink" or as a Cantonese word meaning "genuine". I doubt that they knew that derogative word at their age; most likely they used a Cantonese word as an emphasis. I did not feel offended at that time — I did not grasp the meaning of the term anyway — and neither did my teacher stop them. However, this nickname was dropped after a few months when I adapted myself to the environment. Looking back I would take this as a reflection of the superior feeling of the English-educated Chinese over the Chinese-educated in those days in that part of the world.

More worrying was my acquiring of the habit of stuttering, in the form of a hesitation to read or to say the first word in class. That was perhaps due to my inferiority complex resulting from deficiency in oral English. Perhaps another cause was the first cultural conflict that I encountered around 1934 and 1935. My textbook for Bible History clearly stated 4004 BCE as the year of the Creation and gave dates in the Old Testament that showed discrepancies with the traditional Chinese historical dates given in Chinese texts. I did not question the authenticity of either of them, but could not come to terms with two conflicting views. I hesitated to make a decision — I could doubt neither what my Bible History said nor the traditional dates given in Chinese texts. In the Western tradition as taught in school one had to declare one's choice to the exclusion of any other, while in the traditional Chinese way declaration and rejection were unnecessary. Time has shown that the dates in my Bible History textbook and the early traditional Chinese dates were both unreliable, but perhaps hesitation to commit to the Western tradition in this issue had played a part in the development of my stuttering. I tried hard to overcome this defect, but it persisted for some years and only gradually faded away towards my last period in school. I could hardly imagine then, that later on in life, I could manage to speak to audiences of over half a thousand without reading from my notes.

There were three examinations in a year. At the first term examination I came out slightly below average, 24th out of a class of 40, at the second I came out 18th, and at the final I improved again on the score, but had no recollection of what it was. My examination results kept on improving gradually in 1935 when I was in the Standard Four class, but what left for me to remember was a chain-smoking teacher, who kept himself otherwise occupied whenever there was a horse race in town. I went to Standard Five class in 1936. That was my last year of primary education. I had a very dedicated teacher, Mr Lip Seng Onn 聶成安 as my form teacher. Schooling had become a more serious discipline. I had to learn two new subjects, algebra and geometry, taught by a competent teacher, Mr Lim Kean Hooi 林乾辉. My examination results also improved and came within the first ten in class. In 1937 Mr Lip was promoted together with his class to become the form teacher of Standard Six, the first year in secondary education.[7] My position in class kept on improving. The last year of primary education and the first year of secondary education were my years of consolidation in school. I became the top boy in class in my last three years in school. By that time I had given up my early morning *kung fu* session to concentrate on my homework. The form teacher of Standard Seven class was a German Christian Brother, Brother Rupert. He appeared to me as the most learned Christian Brother during my time at school and he was very dedicated to teaching as well. Among the subjects he taught was French. One day he taught the word *le professeur*. Suddenly my classmates applied that term to me and called me professor. Even Brother Rupert joined them. I was stuck with this nickname throughout the rest of my school years.

---

[7] There were some confusion about Chinese names in the schools in those days. Pupils were taught to address their teachers as Master, followed by their personal names, such as Master Seng Onn and Master Kean Hooi in the case of the two teachers mentioned here. In another school in the same town they would be addressed as Mr Lip and Mr Lim respectively.

I had Mr Lip as form teacher for the third time in the Standard Eight class. My French teacher in that class was an Irish American Christian Brother by the name Brother Thomas. His honesty discouraged me from venturing to speak French. He related to the class that when he once spoke French in France, he was told that he was speaking the language like a Spanish cow. I feared that my performance would at best be equal to that of a Spanish calf. However, I remember what he taught us about thinking big and the American belief in speed and efficiency. That was my early encounter with an American.

War started in Europe in 1939. I attended first aid classes and joined the St. John Ambulance Brigade. 1939 was also the last year of the Overseas Cambridge Junior School Certificate Examinations. That year I also started to learn to play the violin. The music was mainly Scottish. For example, some of the early pieces I learned to play included *Blue Bells of Scotland* and *Coming thro' the rye*. I had not succeeded in becoming competent in playing the violin, although I did once played in the school concert under duress and after persuading the classmate on the piano to play louder to cover up my wrong notes. I did not know whether he did so, but I survived the ordeal. This classmate of mine came from a family of musicians from Goa; he later worked as a piano teacher in Singapore. I stopped my violin lessons in my last year in school to prepare for the Cambridge School Certificate Examinations, hoping to resume such activities after finishing school. However, it was not to be as later event showed. What I have learned was that the violin was not an easy musical instrument to play. An easier musical instrument for me to learn was the harmonica. I used to play pieces of Chinese as well as Western music with it, but like the violin, I had to give it up eventually.

Standard Nine was known as the Senior Cambridge Class. My form teacher was an English Christian Brother, who hailed from Salisbury. My other teachers told the class that he was hard of hearing and therefore was able to keep his pure English accent without being polluted by listening to local people speaking his language. The class was thus advised to listen carefully to him in order to pick up the proper English accent. He was supposed to teach English literature and botany, but he used up most of the teaching hours either preaching or talking about ethics to prepare us for life after leaving school. One of my classmates, Anthony Yong later joined the priesthood and ultimately rose to the rank of an Archbishop. Perhaps it was partly due to the effort of this teacher. A Burmese Christian Brother with the same religious name as our form teacher taught us history of the British Empire. At his last class before we left school, he wrote five history questions on the blackboard and asked us to study them. Sure enough four of those questions reappeared in the history paper of the Cambridge School Certificate Examinations. He must have been an expert analyst of the ten-year series history questions. Most of the Christian Brothers at that time came from Ireland, while those from Asia originated in Burma. Mr Lip taught geography, while Mr Lim taught elementary mathematics, which included arithmetic, algebra and geometry. In December 1940 I finished school and sat for the Cambridge School Certificate Examinations.

I got on very well with all my teachers in school. Most of them were nice people, but I had no idea at all about teachers in other schools. Some of my teachers imparted knowledge that involved many things outside the school syllabus, such as in mundane affairs and religious denominations other than their own. Perhaps with the exception of Brother Rupert, who seemed to have expertise in chemistry, none of my teachers had gone through tertiary education. Only their dedication to education somewhat made up for the lack of it. Certainly, tertiary education was not a common qualification possessed by many teachers in those days. However, the other two English schools in town did have several college graduates as teachers. For example, Mr Lim was a good mathematics teacher and he taught his class elementary mathematics, but the Anderson School had a college graduate mathematics teacher, who maintained a 100% examination success record in his teaching career and who taught not only elementary but also additional mathematics. The other English schools in town also made more effort to prepare students for further studies than the school I attended. They offered additional mathematics, physics and chemistry, which were absent among the subjects taught in my school. Later on I had to make extra efforts in college to make up what I missed. Nevertheless I have never regretted having attended that school and I kept in close contact with it after leaving. It did take steps to remedy this situation in the post war years. In 1971 I served on the Board of Governors at St. John's Institution in Kuala Lumpur and was pleased to note that no less than 40 percent of its teaching staff had received tertiary education. This was a far cry from the days when I was in one of its sister schools.

I got along well with my classmates, although I did not spend much time playing or going out with them. They probably regarded me as a bookworm through my absence in sports and on the play field. The only exception was Choong Sin Choon, who used to visit me to have a game of billiards on a makeshift table and to talk about school homework. After leaving school I was in contact with only a small number of them. When I taught temporarily in my old school shortly after the war, I had Lee Kam Wai 李金辉, Cheah Hoong Tuck 谢鸿德, Ooi Eu Chong 黄耀宗 and Chan Kok Soo 陈国书 as colleagues. Kok Soo was a nephew of Chen Pi-chun 陈壁君, the wife of Wang Ching-wei 汪精卫, the President of the so-called Puppet Nanking Government, but political events in China had no repercussions within the walls of an English school. After leaving my hometown Ipoh, I was in close touch with Lee Kam Wai, who graduated with a degree in medicine at the University of Hong Kong and practised medicine in Singapore. Lucy is a younger sister of his wife, and through him we first met in Hong Kong. Cheah Hoong Tuck later became a colleague of mine at the University of Malaya, Kuala Lumpur, where he worked as the Assistant Bursar. He later moved to Penang as the Bursar of a new university before migrating to Australia to settle down in Melbourne. I enjoyed many years of close friend-ship with them while I was in Singapore and in Kuala Lumpur respectively. I was with Anthony Yong several times at the National Consultative Council meetings held at Parliament House in Kuala Lumpur when he was the Bishop of Penang, and had once invited him to my home for lunch during the interval. I came across Lim Thiam Loy, Ooi Boon Seng and Chang Kong Fee when I was in Singapore. In 2001 when I revisited Ipoh I met Wong Hon Choong 黄汉忠, Lau Yew Meng 刘耀明 and Yeoh Kean Teik 杨建德 for the first time after sixty long years. Kean Teik has a famous daughter Michelle Yeoh 杨紫琼, who starred in one of the James Bond films. Ooi Eu Chong invited me to lunch at the Ipoh Swimming Club, but two years later he passed away.

Interregnum Period: World War II

After I left school towards the end of 1940, my father tried to get me a job through someone he knew. I was supposed to work as a clerk in a local bank without having to submit an application. I heard no news of the result, except to learn from my father that the job would carry a salary of $30 a month. At that time my father had built a house in Tebing Tinggi, a suburban village about two miles away from home and there he established a branch of his school. He taught at the main campus in the morning and at the branch in the afternoon. He got me registered as an untrained teacher to teach English at the branch in the morning and at the main campus in the afternoon. Another full time teacher was employed at the main campus since there were two classrooms. I received no salary for my service, as I was merely working for the family and not being employed. I gave private tuition to earn some pocket money. I did not make any complaint, but I resolved that such an arrangement should not be allowed to last long. I was waiting for an opportunity to further my studies. I applied for a scholarship to enter Raffles College in Singapore, with Dr Wu Lien-teh 伍连德 acting as one of my two referees, the other being a public notary. Dr Wu was the world-renowned plague fighter during the first decade of the last century. With an MD degree from Cambridge University, he was the person holding the highest academic degree that I knew in my youth until I went to university. Strong recommendations alone did not work. The College told me that I was underaged and could not be considered for admission. I was then only 15. I was also hoping to have a chance to apply for a Loke Yew 陆佑 Scholarship to get into the University of Hong Kong to do engineering, but the opportunity never came — there was no announcements calling for applications. I later learned that it was not offered annually.[8] I could not expect any help from my parents since they themselves needed my help; even studying on a scholarship would deprive them of my financial support. I

---

[8] I could never dreamt that 40 years later I was offered the Chair of Chinese in that university with my office on the top floor of the building donated by and named after Loke Yew.

decided to take a correspondence course with the British Institute of Engineering Technology in London to study for the Institute of Mechanical Engineering examinations. I was not aware of the requirements of practical experience at some stage. As events turned out this problem proved to be inconsequential, but I did learn something about differential and integral calculus in the practical mathematics course that became helpful when I later took mathematics in college. That correspondence school soon set up a branch in Sydney and operated from there. That was my earliest contact with Australia. That was the year 1941. People sensed that war was coming near to the country. I enrolled myself to serve in the Medical Auxiliary Service as a member of the St. John's Ambulance Brigade and attended regular training sessions. I did rather well in the tests, scoring full marks for my written test and having two marks deducted from my practical test for failing to correct a mistake deliberately made by the examiner when he assisted me to lift a patient from the stretcher. He said later that it was to make it more difficult for me to gain full marks in both written and practical tests. The study of first aid did make me interested in medicine, but I had to suppress the thought of reading medicine due to my responsibility of supporting my parents. To do medicine over six years was too long to be considered.

The Pacific War began on the 8th of December 1941. My parents, myself, and my brothers and sisters moved to the branch school in Tebing Tinggi. Public defence units were put on alert. The Medical Auxiliary Service had the Town Hall as its headquarters. The unit was divided into two shifts, rotating to report for duty. There were several air raids by Japanese planes and some small bombs were dropped in the town. There were some casualties among the civilians, but only a few houses were damaged. It so happened that the air raids came only during periods when I was off duty. I did not have to attend to the wounded and had not even seen any blood throughout the period of my service. One December morning when I reported for duty I found the Town Hall deserted. On making enquiry, I found that those on shift duty last night had evacuated to the south and were on their way to Singapore in the face of the approaching Japanese enemy, leaving those in my shift behind. That ended my career in the St.

John's Ambulance Brigade. After the war I received a Defence Medal bearing the image of the head of King George VI for service unaccompanied by performance, I felt rather undeserving, I could offer no reason to justify myself.[9]

The Japanese army gained control of the Malay Peninsula very quickly. It entered Ipoh before Christmas and occupied Singapore on 15th of February 1942. Ipoh was relatively peaceful, probably because the Japanese met little resistance on its way. Under Japanese occupation some English schools were turned into Japanese schools, while some were used for other purposes. St. Michael's Institution became the headquarters of the state government and the Convent School was turned into the Perak Japanese Language Teachers Training Institute, which was later renamed Perak Normal School. It offered a Rapid Course lasting three months for existing teachers and a Specialised Course that stretched over six months for new recruits. I joined the first batch of Specialised Course students in January 1943. I put in much effort in studying the language. I copied a whole Japanese-English Dictionary by hand, as dictionaries and English-Japanese language books were rare items. I also copied a portion of a Japanese language book by Chamberlain and later I had a brief access to Vaccari's first edition of *Japanese Conversation Grammar*. Of course Japanese language in those days was not taught via the English medium, but the English educated found it extremely useful. The Head of the institute was also the Head of the Education Department. The first Head was Shijiro Banno 伴野知志郎, a graduate of Waseda University, if not Michigan University. He compiled a textbook on Japanese grammar for use in the institute. My knowledge of classical Chinese complemented the English grammar I learned in school to help me in the study of the Japanese language. Lessons taught in the institute came from textbooks used in primary schools in Japan. All classes, including the grammar classes taught by Banno himself, were conducted in the Japanese medium. In May 1943 Banno was transferred to the post of Head of the Propaganda Department. A new Head for the institute and the Education Department arrived in the person of Hiroshi

---

[9] This came about in 1947. I kept the medal as a souvenir, but I also received a back pay of about $30, which was a windfall to me in my student days.

Suguro 胜吕弘, a graduate of Tokyo University of Commerce (former name of present Hitotsubashi University) and a professor in a Japanese university with expertise in marine insurance before he was recruited as a military administration officer. In end June 1943, I came out top of the class in the graduation examination. I was appointed to serve as a Teaching Assistant in the institute and also as a clerk in the Education Department situated at the second floor of the Hongkong and Shanghai Bank building in Belfield Street.[10] In 1944 I was transferred to the main government department located in the same building as my *alma mater* to work as a clerk in the Administration Department.

The Pacific War ended in August 1945. It did not take long for the English schools to reopen. I was offered a teaching post in my old school and was in charge of organising classes to get them ready for the established teachers to take over. I found myself teaching several classes within the period of a few months. My school also asked me to take on the after-school Chinese classes. I had to do something about my ability in *putonghua* 普通话, popularly known as Mandarin. I came from a Cantonese speaking family. My encounter with Mandarin came only from two sources, listening to Chinese songs and to Mr Wang Fo-wen 王宓文 talking to my father. Mr Wang, the father of Professor Wang Gungwu 王赓武, was the Assistant Inspector of Chinese Schools in the State of Perak then, and he was a close friend of my father.[11] Ipoh had a large population of Hakka people and some took the Hakka dialect as *putonghua*. To acquire a correct pronunciation in

---

[10] The bank, of course, was not functioning then.

[11] My father spoke to him in *guanhua* 官话, the sort of language used by Chinese mandarins, but it was not *putonghua*. Wang Fo-wen and his family stayed with my family in my father's house in Tebing Tinggi at the initial period of Japanese occupation. We used to spend our time practising Chinese calligraphy on old newspaper. Wang Fo-wen was an expert in the seal script type of calligraphy. He made his son Gungwu practise the Yan 颜 style of writing, while my father made me learn the Zhao 赵 style. Gungwu and my younger brother Yew Lam were in the same class at the Anderson School, Ipoh. I gave them help in mathematics to prepare them for their school education. Mrs Wang and my mother were great friends; she picked up some Cantonese from my mother, while my mother was unable to speak in *putonghua* at all.

putonghua, I got a helping hand from Lee Weng Toon to learn the Chinese phonetics known as *zhuyin fuhao* 注音符号. I had to check every word in the lesson before facing my class.

Things were again not the same for my father. My meagre salary and that from my brother Yew Lam, who was also working as a clerk, were barely sufficient to meet his needs. He sold his small house in Tebing Tinggi, which was the only real estate property he acquired in his lifetime. Eventually he leased out No. 30 Theatre Street to a sundry shop owner to live at No. 28 Theatre Street to get more income. When war first came we lived in Tebing Tinggi, and my father invited his elder sister and her children to live in No. 30 Theatre Street as she had lost her husband. My personal belongings, such as school reports and books, stamp collection and music books were kept in my father's study at the same premises. I suppose that what's one's treasure can be somebody's garbage, to borrow an often repeated statement of my wife Lucy. The loss of my old possession discouraged me from starting to collect stamps again, although I casually save my used stamps to be given away. My loss was insignificant and merely had some sentimental value, but my father had to cut down his school to one-third its former size in terms of number of classrooms. He operated his school with only one classroom until the very last day of his life.

My experience during the war years opened my eyes to the world beyond the walls of my old school for the first time. I regarded such experience as an important part of my education, although I seldom spoke about it. At the Japanese Language Teachers Training Institute I helped to teach many school-masters from English, Malay, Chinese and Tamil schools in the State of Perak. Among them were my former teachers. I was also assigned to hold Japanese class in town for the public to attend. Some teachers and even some Chris-tian Brothers from my *alma mater* attended the classes. I made friends with many teachers from other schools, besides getting to know some of my former teachers better. I observed that there were dedicated and good teach-ers in other schools just as there were in my *alma mater* and that there were nice people in other Christian denominations just as there were among those of my *alma mater*. My relationship with some of the Christian Brothers and my former teachers became even closer. Mr Lim, my former mathematics teacher, became my fellow Teaching Assistant in the institute. When Mr Lip studied at the institute, he was very pleased to find me there ready to help him with his Japanese. My contact with teachers from Chinese, Malay and Tamil schools showed me some cultural differences between the English-educated and those educated in other language media.

Even more important was the development of a happy relationship with the Malays. For example, among those working together as Teaching Assistants in the institute was Hamdan, who was later known at the height of his career as Tun Dato' Seri (Dr) Haji Hamdan Sheik Tahir.[12] He served as Director of Education of Malaysia and Vice-Chancellor of the University of Science of Malaysia in Penang before he became the Head of State of Penang (Yang Dipertua Negeri Pulau Pinang), known at one time as Governor. I became his houseguest at Seri Mutiara, the Residency of the Head of State whenever I visited Penang during his term of office. Our friendship has been so close that we regard each other as brothers. His spouse Toh Puan Siti Zainab, still refer to me as Mr Ho, because I once taught her Japanese. Another friend, Tan Sri Raja Mohar bin Raja Badiozaman, later became Director-General of Federal Treasury and served as financial advisor to the Prime Minister. He was Chairman of the Malaysian Airways System (MAS), while I was Chairman of the School of Modern Asian Studies (MAS) at Griffith University in Australia. He offered free travel in first class on his airline to Malaysian scholars invited by me to visit Australian universities for the promotion of mutual understanding between scholars from Malaysia and Australia.

At the Administration Department I came to know many Malay civil servants, including District Officers and Malay Officers, as they needed my help to communicate with Japanese officials. Among the civil servants I came to know were Mustapha Albakri, who later served as State Secretary to Perak and Commissioner of Election. He was Chairman of the University of Malaya Council when I was the Dean of Arts at the same university in

---

[12] Tun and Dato' Seri are both Malaysian civil titles. Tun is the highest title awarded by the Head of State of Malaysia. There is also the tile Tan Sri, something like a knighthood that comes after Tun. Hamdan prefers to indicate his honorary doctoral degrees within brackets.

1968.[13] I remember both of us once attended a sub-committee meeting chaired by the Vice-Chancellor. There was only one other member at the meeting and he was the Dean of Science, who also happened to come from Ipoh. The Vice-Chancellor, Dr James Griffiths, was a Fellow of Magdalen College, Oxford. Mustapha Albakri revealed to Dr Griffiths that all the three members attending his meeting were from Ipoh. Dr Griffiths, with his sense of humour, started the procedure with the remark, "I surrender". Bahaman bin Samsudin, who was then a District Officer transferred to the Administration Department under suspicion by the Japanese for helping the British Force 136, gave me advice on two things in life — never ride on a motorcycle and try to avoid becoming a schoolmaster. The reason for keeping away from motorcycles is risk. He said that a schoolmaster usually did not have many friends after acquiring the habit of talking down to school children and often unconsciously adopted the same stance towards others. I took his first advice to the letter, but interpreted the second as an advice of never to adopt the stance of a teacher outside the classroom. Bahaman later became the Minister of Health and a member of the University of Malaya Council when I was Head of the Department of Chinese at that university. Dato Osman bin Talib, another District Officer I knew, later became the State Secretary of Perak and later Chairman of the National Electricity Board.

Sultan Abdul Aziz of Perak also called for my assistance, not only to act as interpreter for him when he spoke to the Japanese Governor, but also to help the British army take things over from the Japanese military after its surrender. The unit I helped belonged to the Berkshire Yeomanry. I once had lunch with Sultan Aziz at his palace in Kuala Kangsar. I was also asked to join him as his interpreter during one of his visits to Grik in Upper Perak. His successor was Sultan Yusof Izzudin Shah, who often referred to me as his young friend, although I was actually his subject. In those days people born in the Straits Settlements, like Singapore and Penang, were British subjects and those born in the Malay States, such as Perak, were subjects of the

---

[13] I was confused by the change of street names in Ipoh during my visit in 2001. I felt somewhat relieved when I noticed a street nearby my former home was renamed after him.

Sultans and only had the status of British Protected Persons in their passports. Once, I happened to be travelling in the same train as Sultan Yusof from Singapore to Kuala Lumpur and he invited me to the royal coach to join him for lunch. In October 1949 at the University of Malaya Foundation Day ceremony held in Singapore, he introduced me as a friend of his to the Sultan of Pahang and told me that the Sultan was a relative of his by marriage.

I was also very lucky with the Japanese I came in contact with. They were quite different from those in the military. Those who came to serve in the Japanese Language Teachers Training Institute were teachers in Japan and they developed friendly relations with many local people that lasted long after the war. Towards the end of 1943 Saburo Kodera 小寺三郎 succeeded Hiroshi Suguro in the institute as well as in the Education Department when the latter gained promotion to work in Singapore as Education Chief of the whole of Malaya. Kodera was a graduate in English Literature from Tokyo University. His mannerism and fluency in English made some of my fellow Teaching Assistants regard him as an English gentleman. He sent his Japanese colleagues to serve all over the State of Perak and advised them not to carry their swords along with them, in order to show that they were men of peace. None of them came under attack by anti-Japanese resistance forces hiding in the jungle, while other Japanese continued to suffer casualties. There was some suspicion that he had made some sort of deal privately with the resistance forces, but no evidence emerged. All that was done to him was to make him live with the Governor in his Residence to keep him away from mischief. Officially he acted as the Governor's interpreter, while keeping his post as Education Chief. When the Japanese surrendered he played an important role acting as interpreter for both sides until he was finally repatriated to Japan. He married into the Aisawa family of Okayama, a family of politicians and proprietor of a large construction firm that built many roads and bridges in the Kansai and Chugoku area, including part of the bridge connecting Shigoku Island to the mainland. He himself had once contested a seat in parliament, but was unsuccessful. I think his gentle manner acquired from being a former English teacher in a women college was not suitable for him to become a politician. He then became a senior executive officer in the Aisawa Construction Company and a member in its Board of Directors. His brother-in-laws and his nephews carried on the family tradition in parliament. He was always very busy during election periods when he had to join the Aisawa family members to render a helping hand to their candidates.

When I was teaching in school, I met with an accident while playing volleyball with the students. I stepped on a pile of sand just behind the court and fell with a dislocated left elbow. It was not properly fixed and I was left to live with an elbow that could not be fully bent. I have never spoken about this accident outside my family. My fingers could no longer reach the full extent of the violin fingers board, and I had to give up learning to play that musical instrument altogether. However, this accident saved me from the trouble of having to make a hard decision at a later stage.

From College to University

The Raffles College and the King Edward VII College of Medicine announced their reopening and called for applications for new admissions. I applied for a government scholarship to enter Raffles College, again with Dr Wu Lien-teh as one of my two referees. I had become better acquainted with Dr Wu by then. Our first meeting in 1941 was through his younger brother Ng Tuck Onn 伍德安, whose son Cheong Fok 长福 attended evening class at my father's school to learn Chinese. I met Dr Wu on several occasions during the war years and after the war both of his two sons, Chang-sheng 长生 (Fred) and Chang-yuan 长员 (John) were in my class when I was teaching in my *alma mater*. I was only four years older than Fred and had played badminton together with both Fred and John during after school hours. Fred later took law in Emmanuel College, Cambridge and became a prominent barrister in Singapore specialising in insurance, construction and arbitration laws. Dr Wu introduced his eldest daughter personally to me, saying, "Meet Betty, my daughter." Betty, better known in Chinese by her name Yu-lin 玉玲, made a great contribution to Singapore and the ASEAN countries in education by serving as the first Director of the Regional English Language Centre in Singapore.

My second attempt to join Raffles College had better luck. I gained admission to the College with an award of a government scholarship. I heard that a list arranging the candidates in order of merit strictly according to the Cambridge Senior School Certificate examination results of 1940 and 1941 was in the hands of both colleges and that I could have obtained a scholarship to take medicine had I applied. I did not regret my choice, not because of preference of science to medicine, but because my parents needed my financial support. Three years in science were long enough for them to wait. In those days medicine was the first choice for students in Southeast Asia undertaking higher education. The second choice was law. To the English educated monetary reward and independence were the common considerations for the choice. Traditional Chinese shared the same view with the English educated regarding medicine, and educated Chinese would refer to an old saying that if one does not become a good prime minister one should become a good doctor — *bu wei liang xiang ze wei liang yi* 不为良相则为良医 — regarding medicine as a noble profession.[14] However, traditional Chinese did not accord the same esteem to law, believing that lawyers would help wrong-doers escape justice and influence the court to pass wrong judgement in civil and criminal proceedings.[15] In any case taking up law was not within my reach and was not in my mind either. I selected to study science from two choices available to me, the other being the humanities.

Friends in need are friends indeed. My scholarship would cover only college tuition and boarding fees, with hardly anything left for buying books and pocket money. Having deprived my parents of income derived from my school salary, I could not depend on them for assistance in this matter. Fortunately along came two saviours. Dr Wu Lien-teh and Mr Lip Seng Onn both asked me to write to them when I needed money. As my intention was only to borrow, I thought that Dr Wu would insist that his money was

---

[14] Four of my five children took up medicine as their career. My son is Professor of Surgery at James Cook University, Queensland. Only one daughter followed my career in science to become a specialist in aerospace dynamics.

[15] Uncle Qixun told me that he had never lost a single case when he was practising law in Canton, but quickly added that he only accepted civil cases, perhaps to mitigate my opinion of lawyers, as he was only taking on the lesser of the two evils.

a gift and not a loan. Mr Lip did not say whether it would be a loan or a gift, but it would be much easier to talk to him over the matter than to talk to Dr Wu. Altogether I had borrowed from Mr Lip a sum of $160 throughout my three years in college. I returned the full sum to him immediately after I got my salary from teaching in my *alma mater* upon graduation. I earned some pocket money by given private tuition during my college vacations, but sometimes it was *gratis* when I helped Mr Lip's eldest daughter Mary Michael to learn mathematics.

On hearing of my admission to Raffles College, Kok Ah Loy 郭亚来 from another school in Ipoh came to see me. He said that he was studying science at Raffles College before the war, but because he wasted too much time on his violin that he failed his examinations. I appreciated the friendship and frankness of this friend, who later proved that his failure in college was not due to lack of ability by obtaining a B.A. (Special) degree from London University as an external candidate. My fall resulting from playing volleyball had already made my decision for me. Nevertheless I took heed of his words, not allowing myself to digress during my college career. Knowing how good a student he was in school, I had to be very careful to avoid what happened to my friend.

In October 1946 I entered Raffles College as the holder of a Malayan Union Scholarship and a First Year science student. I had to take four subjects, namely, Pure Mathematics, Applied Mathematics, Chemistry and Physics. In my Second Year my scholarship changed its name to Federal Scholarship following rapid political changes in the country. Second Year students had to drop one of the four subjects that they took in the First Year. It was a difficult choice for me. My decision to drop Chemistry was based on the fact that no professor was appointed yet for that subject. A new professor was appointed for Chemistry only when I was midway through the Second Year, but by then it was too late for me to think about Chemistry. Remembering a saying among schoolmasters who received their education in China, *du liao shu li hua, keyi zoupian tianxia* 读了数理化可以走遍天下 — having learned (one of the subjects of) mathematics, physics and chemistry, one can go anywhere in the country (and find employment) — I thought that any of these three subjects would be good for me.

I was elected table tennis captain in the Students Athletic Committee and Auditor of the Students Union. A notable event was Dr Wu Lien-teh's visit to Singapore to give a talk at the invitation of the science students' body, the Physical and Chemical Society. Attending his talk was the grand old man of Singapore Dr Lim Boon Keng 林文庆. Both Dr Wu and Dr Lim were Queen's Scholars during their times in the 19th century and they married two Wong sisters. Dr Lim obtained his medical degree at the University of Edinburgh and Dr Wu in Cambridge.[16] Dr Lim had served as the Foundation President of Amoy University. He was then approaching his 80th year and everybody was amazed to see him walking up the stairs unaided. Dr Wu introduced me to him, and that was the first and only occasion that I met this grand old man of Singapore.

I organised a science display at Raffles College together with my fellow students and attracted good attendance from the public. During my Third Year I repeated the function with full encouragement from the teaching staff. The name of the event was changed to Science Exhibition. While organising the exhibition I came into contact with a number of industrialists in Singapore. From them I heard about the shortage of science graduates in their employment. I gave a talk on Einstein's Theory of Special Relativity on behalf of the Physical and Chemical Society. In sports I was elected table tennis captain for a second year and also won the Oppenheim Shield as the College Chess Champion of the year. Meanwhile Professor Alexander Oppenheim gave tuition to Thong Saw Pak 汤寿柏 and me at his office on homogeneous geometry, a topic beyond my Third Year mathematics syllabus.

I thought that my academic career would end after gaining my Raffles College Diploma in 1949 and that I would be looking for a job either in a school or in the private sector in Singapore. I expected that my *alma mater* would have a place for me and I also heard through my contact with the

---

[16] An early reference was given by Song Ong Siang in 1922 in his *One Hundred Years' History of the Chinese in Singapore* (reprinted 1967, University of Malaya Press, Kuala Lumpur and Singapore). Song was another Queen's Scholar, but he took up law in Cambridge. Lim and Song were prominent Chinese in Singapore some two generations before my time.

private sectors in Singapore during my organisation of the science exhibition in the College that there were a number of vacancies for science graduates in the industry. But the University of Malaya was to be founded that year and Professor Norman Alexander approached me before I left the College and asked whether I would like to become a Physics Honours student in the new university. I was only too delighted to accept his offer, and that was the main reason why I took up Physics.

The name of the University of Malaya requires clarification because of its occurrence in different places and times. The word "Malaya" was a geographical name that included the Federated and non-Federated Malay States together with the Straits Settlements before the Second World War. Singapore, being the seat of the British Governor, was the most important member of the Straits Settlements as well as of Malaya. It was natural for the highest seat of learning of the country to be situated there. After the war Singapore obtained self-rule, while the rest of Malaya did likewise under the name "Federation of Malaya". They were not yet independent entities and geographically they were still known together as Malaya.[17] It was quite appropriate to choose the name University of Malaya when Raffles College amalgamated with the King Edward VII College of Medicine in 1949 to form a new university. Nearly a decade later, the Federation of Malaya decided to have the seat of higher learning in its capital, Kuala Lumpur. The University of Malaya split into two divisions, one known as the University of Malaya in Singapore at its original campus and the other as University of Malaya in Kuala Lumpur, situated in Pantai Valley between Kuala Lumpur and its satellite town Petaling Jaya. Each division had its own Principal, but they shared a common Vice-Chancellor. Some members of the teaching staff from Singapore were transferred to Kuala Lumpur. Eventually the two divisions decided to become totally independent and have their own Vice-

---

[17] In 1962 I went to a post office in New York to send a registered letter to Singapore. I was asked where Singapore was. Upon supplying the longitude and latitude of Singapore, the postal clerk said that he only wanted to know which country Singapore was in. He then looked up a book and showed me that the postal location of Singapore came under the country called "Malaya".

Chancellors. In 1962 the Kuala Lumpur division retained the name University of Malaya, while that in Singapore became the University of Singapore. The next year, the Federation of Malaya and Singapore together with Sarawak and former British North Borneo, now known as Sabah, formed an independent country known as Malaysia. Singapore separated from Malaysia to become an independent country in 1966. In the early 1970s the University of Singapore merged with the Nanyang University to become the National University of Singapore.

Coming back to the time I left college, I belonged to the last batch of graduates produced by Raffles College. In a way, together with my fellow graduates we had lost four or five precious years due to the war and entered college at an age when a pre-war student would have already graduated. On the other hand we felt ourselves more mature. The same applied to the next few batches of college and university student intakes after us. These few batches of graduates came at the right time though. As a result of independence, many graduates filled up the vacuum created by expatriate civil servants leaving the country. Even foreign establishments in the private sector sought after local graduates. That was quite different from the pre-war days when most of the Raffles College graduates became schoolmasters. Hence among my contemporaries in college I could count a future king (Yang Di Pertuan Agong) of Malaysia and future cabinet ministers, senior civil servants, academics and industrialists besides schoolmasters. Meanwhile the King Edward VII College of Medicine produced not only doctors and dental surgeons but also politicians, one of them later became Prime Minister of Malaysia. These early batches of graduates were indeed given a good start in life and few of them ever encountered the ordeal of unemployment. I considered that we were extremely fortunate when I reflected on the plight of university graduates in pre-war China who interpreted graduation as unemployment — *biye jiushi shiye* 毕业就是失业. We were at the right place at the right time.

I returned to Ipoh to teach in St. Michael's Institution as a temporary science teacher for three months. The school was hoping that I would return to teach there after I obtained my Honours Degree. It was in need of science graduate teachers. Graduates and undergraduates of the two Singapore colleges organised a dance to raise funds for the University of Malaya Endowment Fund. I was elected Chairman of the fundraising committee. We invited the most important dignitaries in the State of Perak, including the Sultan of Perak, the British Advisor, the Mentri Besar (State Chief Minister) and the Consul of the Republic of China (Ma Tien-ying 马天英) to be Patrons. I went with my committee members to every house in town to sell tickets. The dance was held at the Jubilee and the Celestial Cabarets, the only two dance halls in Ipoh in those days. Our Patrons also showed up. I was sitting next to Sultan Yusof and I noticed the smile on his face. The sum of money we raised reached five digits, which was a record for similar efforts made by graduates and undergraduates in other parts of the country. Figure 2 reproduces the list of committee members as shown in the original fundraising programme.

I returned to Singapore in October 1949 to enrol myself as an Honours student. The university made arrangements to have my Federal Scholarship extended for another year. Professor Alexander gave me a part-time job as a student demonstrator in the laboratory classes. There was no longer any need for me to borrow money to supplement my scholarship. After I gained

---

Members of the Varsity Dance Committee.

Chairman:　　Mr. Ho Peng Yoke (Raffles College, Faculty of Science).

Secretary:　　Miss Lam Lai Cheng (Raffles College, Faculty of Science).

Treasurer:　　Mr. Ong Cheng Hooi (College of Medicine).

Miss Lam Ah Ngan (Raffles College, Faculty of Science).

Miss Betty Wu (Raffles College, Faculty of Arts).

Mr. Yen Poh Loke (College of Medicine).

Mr. Ng Mann Loke (College of Medicine).

Miss Yohn Khong (College of Medicine).

　　This Varsity Dance is the first of its kind whereby the students of Raffles College and College of Medicine have synchronized their efforts in an all-out attempt to swell the University of Malaya Endowment Fund.

Figure 2 Committee members of the Varsity Dance, 1949

First Class Honours in my degree examination, Professor Alexander asked whether I would like to become a Demonstrator in Physics and do a Master Degree at the same time. I jumped at the offer, with the hope that I could make physics my career. At that time those who obtained a First Class Honours Degree were usually successful when they applied for the Queen's Scholarship to further their studies in Britain. My circumstances would not permit me to take that into consideration, thus Professor Alexander's offer gave me the best I could wish for.

In those days there were not many students doing science and the passing rate of yearly and final examinations was usually no better than 50 percent. In my time there were about 20 students in the second year and about 10 in the third and final year taking physics as a subject. Student numbers for mathematics were slightly larger, because they included students from the Faculty of Arts. There were then only two professors in the whole Faculty of Science, namely Professor Alexander of the Department of Physics and Professor Oppenheim of the Department of Mathematics. I think that several things I did during my student days must have drawn the attention of my two professors. The science display and the science exhibition at the College received much support from the teaching staff and attracted some publicity in the local newspapers. Both the two professors came to my talk on the Theory of Special Relativity organised by the Physical and Chemical Society. Professor Oppenheim gave me extra-curricular tuition, while Professor Alexander took me out for fieldwork to assist his wife, Dr Elizabeth Alexander to do a geo-potential survey of Singapore. It was Dr Elizabeth Alexander who received, as Secretary to the University of Malaya, the cheque for the University Endowment Fund that I sent from Ipoh. Professor Oppenheim personally handed over to me the Oppenheim Shield that I won in 1949 as the Raffles College Chess Champion. My choice of physics was just by chance. I had more opportunity to see Professor Alexander in the physics laboratory than Professor Oppenheim at the Department of Mathematics. Had Professor Oppenheim approached me before Professor Alexander, I would have given him the same answer to go for mathematics. However, Professor Oppenheim later became the person who played the most crucial role in changing my career.

I learned a lesson about bureaucracy from the experience of fundraising for the university. My committee was so happy over what it had achieved that it would like its donation to be placed as a single entry on record. I wrote to Dr Elizabeth Alexander and she agreed with my proposal. She was then doing the job of a university registrar, but a Registrar was not yet appointed pending the official foundation of the university. She was a brilliant student in Cambridge and held a doctoral degree in geology. However, someone else was appointed Registrar instead. She must have been quite disappointed, and this was probably one of the reasons why Professor Alexander left Singapore a few years later. I went to see the new Registrar when I noticed our Ipoh donation missing in the official list of donors. He said for the sake of tidying things up he had lumped the donations from Ipoh together with donations from other graduates and undergraduates. I thought that it would be indiscreet to invoke the consent given by Dr Elizabeth Alexander, an unsuccessful contender for the post he held. He might have taken action deliberately to tidy up things that were handed over to him. I learned from this episode that, whatever hard work we put in to raise funds could be effortlessly disposed of by a stroke of the pen in the hand of the bureaucrat. It would be futile and counter productive to pick a quarrel with him on this score. I did not dispute his right, but I thought that his action did not help to rekindle any enthusiasm in me to repeat a similar performance that was done only voluntarily and not out of duty.[18]

In 1950 I became Demonstrator in Physics and earned a Shell Research Fellowship for two years to enable me to do a Master Degree. The word "demonstrator" acquired a new meaning in the next decade, but even then it took on more than a single meaning. One late afternoon a friend took me by car to fetch another friend to go for a game of tennis. On arrival at his home

---

[18] I did not bear any grudge against this Registrar, who happened to be an expatriate recruited from Britain. A few years later a local girl married to an expatriate Lecturer in another Faculty attempted to occupy the garage that belonged to my university quarters. Instead of confronting her or her husband, both of whom I knew, I informed this Registrar about the matter. Without even asking me further questions he picked up the telephone and settled the issue once and for all.

a sister of this friend opened the door to let us in. My friend introduced me to her saying, "This is Mr Ho, Demonstrator in Physics at the university." She cast her eyes at me from head to foot and said with a smile, "You don't look like one." I thought that for physics she heard physique — I was then weighing only 52 kilograms. At the early stage of the Physics Department at the University of Malaya, research was focussed in local atmospheric problems. Professor Alexander was helping his wife, Dr Elizabeth Alexander in her geological research of Singapore using the electric potential method. There was a small team of two persons from the D.S.I.R. in Slough conducting measurement on the ionosphere. All the others worked on different aspects of the atmosphere, with Mr Charles Webb as the most senior member. I worked with Mr Webb and completed my Master thesis on indoor air movement within one year. In 1951, I was awarded the Degree of Master of Science and my Shell Research Fellowship terminated with the completion of my research project.

2

Career as a Physicist

In 1951 I was appointed Assistant Lecturer in Physics. I began to do research towards a Ph.D. degree. My first publication appeared the next year in an engineering journal in England.[19] It related human comfort with temperature, humidity and air velocity in terms of a mathematical expression. The numerical values it gave for the three comfort factors were sometimes referred to as the Ho coefficients. However, I soon became uncomfortable with both the comfort factors and the method of research itself. The feeling of comfort is rather subjective, not only does it vary from person to person, it also depends on other factors beyond the three that can be measured, human emotion for example. In this sort of research the degree of human comfort was expressed in numbers. I began to wonder whether it was strictly correct to make mathematical calculations involving numbers that could not be measured scientifically. I had to make the apparatus myself for measuring air movement. I was worried that there was no way to check the accuracy of my work. Once the wind blew over it could not be repeated for me to verify my result. In other words, I had lost direction in my research. In 1952 I was selected for the summer school program at the Massachusetts Institute of Technology to participate in the measuring of cloud particles. Everything was already worked out by the American agency in Singapore, what I was waiting for was the result of my application for a Fulbright Travel Grant

---

[19] Ho, P.Y. (1952), "Correlation of Equatorial Climatic Factors with Comfort", *Journal of the Institute of Heating and Ventilating Engineers*, 204:196–197.

submitted through Singapore. Months later I got the news that Singapore would not forward my application because I was not born there. It was already too late to ask Kuala Lumpur to forward my application for me, and so I wrote to decline the offer from the MIT. I might have found a new direction had I attended the summer school, but on hindsight I doubt that counting cloud particles would offer me much scope in my future research. Shortly afterwards Professor Norman Alexander resigned from the university to take up appointment as Vice-Chancellor in a university in Nigeria. I was left working alone without guidance. Fortunately, in my desperation something else emerged to alter the course of my research quite radically.

First Contact with Joseph Needham

In 1951 the University of Malaya established a Department of Chinese Literature and built up a collection of Chinese books in the University Library. Among the books I found a few on the history of Chinese mathematics and began to read them. It happened that Mathematics Department approached me to give a talk. Thinking that the only thing new to my colleagues in the Mathematics Department that I could talk about was Chinese mathematics, I showed them the Chinese method of solving cubic equations and higher degree numerical equations known in China several centuries before the time of Horner and Ruffini. Professor Oppenheim was among the audience and I could notice his smiling face. It was he who taught me Horner's method in my Third Year in Raffles College, and one of the questions in the Final Year Examination was on Horner's method. I also gave a public talk on some aspects of Chinese science at the university that had attracted the attention of the *Straits Times*.

In 1953 I turned to Professor Oppenheim for advice. He was then Dean of the Faculty of Arts and the Dean of Science was Professor R. A. Robinson. In the words of Dr Rayson Huang 黄丽松:

> The two were great friends and were among the very few on the
> professoriate who enjoyed international repute as scholars. Both

held DSc. degrees from leading universities in Britain, and these were rarities in Malaya in those days.[20]

Professor Oppenheim was also a close friend of Mr Charles Webb, who was Senior Lecturer in the Physics Department and was appointed Departmental Head after the departure of Professor Alexander. They were colleagues before the war and both were taken to Thailand as prisoners-of-war and it was while in detention that Professor Oppenheim wrote a research paper on the theory of numbers. Hence I expected that my action of crossing faculties and departments to seek help would not encounter too many diplomatic and administrative problems.

Professor Oppenheim was very sympathetic with my position and suggested to me to consider changing my field of research to history of Chinese science. I asked if he would be my supervisor. He readily agreed with the proviso that he could find an external expert on history of Chinese science to be my advisor. Without delay he started looking for an external expert. He found two names. The Acting Head of the Department of Chinese Literature, Mr Ho Kuang-chung 贺光中, was about to leave for Japan to purchase books for the Chinese collection in the University Library. Oppenheim asked him to find out some names of Japanese experts on the history of Chinese science. He came back to report that he heard about Professor Kiyosi Yabuuti 薮内清 of Kyoto University, and Professor Oppenheim asked Mr Ho to write to find out more. There was no further development in this matter. However, Dr Rayson Huang came along with the name of Dr Joseph Needham of Cambridge University, mentioning the *Science and Civilisation in China* project and adding that he made Needham's acquaintance during the war in Chongqing, Sichuan province. Oppenheim asked Rayson Huang to write to Cambridge and got a reply from Needham saying that he would be glad to correspond directly with me. The two names given to Oppenheim were indeed those of leading scholars in the world on

---

[20] See Rayson Huang (2000), *A Lifetime in Academia* (Hongkong University Press, Hong Kong) 64.

the subject in the 20th century. It would make things easier when communication was established with only one of them and that turned out to be Cambridge, which Oppenheim would have preferred.[21]

I asked Needham whether I could produce a full translation and study of the work of a 13th-century Chinese mathematician for my doctoral thesis.[22] He replied saying that the topic was unsuitable for him, because he had already completed the mathematics section for volume 3 of his *Science and Civilisation in China* and that what he needed to refer to for the astronomy section that he was then writing was a full translation and annotation of one of the *Astronomical Chapters* in one of the earlier Chinese *Dynastic Histories*. He suggested to me three choices: namely, a full translation of the *Astronomical Chapters* in either the *Jinshu* 晋书 (History of the Jin Dynasty) or the *Suishu* 隋书 (History of the Sui Dynasty), or a full-scale study of the scientific contributions of the 8th-century monk, Yixing 一行. He would then make use of the draft I send him for his own book and at the same time add his comments and suggestions so that I could use them for my thesis. I selected the *Jinshu* and at the same time incorporated material from the *Suishu* in the annotations, because the *Astronomical Chapters* in both were written by the celebrated early Tang astronomer and astrologer, Li Chunfeng 李淳风.

Thus in 1953 I became Oppenheim's one and only Ph.D. student as well as one of the early collaborators in Needham's *Science and Civilisation in China* series, while Needham became my advisor for my Ph.D. thesis. Although Oppenheim did not know any Chinese and neither was he an astronomer nor a historian of astronomy, he made an excellent supervisor. He showed me the unwritten basic rules of writing a thesis and went over every sentence in my draft. He asked questions, suggested what should be further explored and what might be skipped over. This experience later led

---

[21] I later came to know Yabuuti and enjoyed his friendship for 40 years from 1959 to 1999, which was a year before he died. He had never mentioned any letter concerning me before we first met and neither did I ask him.

[22] In the 1960s a Belgian scholar worked on this topic for his doctoral degree. See Ulrich Libbrecht (1973), *Chinese Mathematics in the Thirteenth Century* (Cambridge, Mass.).

me to supervise a few Ph.D. candidates working on areas entirely outside my own. For example, I was obliged to supervise a native Japanese tutor in another Australian university to write a Ph.D. thesis on ancient Japanese history. Of course this sort of venture should not be attempted under normal conditions. It does not always work, unless the candidate is highly motivated and is capable of working independently, and as for the supervisor it is not recommended unless circumstances demand it.

I started working on my thesis as soon as I made my choice, getting Oppenheim to go through my drafts with me each time before I sent them to Cambridge. By 1954 my supervisor and my advisor were able to report to the university that I had made substantial progress towards my Ph.D. thesis. As a result I was appointed Lecturer in Physics. At last I felt some sense of security by having obtained a tenured employment. It was a great relief to me, as I had been struggling quietly to support my parents. In a way 1954 was another starting point in life for me. That year I went to Hong Kong with the university students athletic team for the inter-university games and met my future spouse Lucy Fung (Fung Mei Yiu 冯美瑶), the third daughter of Mr Fung Kean Yu 冯镜如, a prominent businessman in Hong Kong during the second quarter of the 20th century and the chief assistant of his uncle Mr Fung Ping Shan 冯平山, who made major donations to the University of Hong Kong in the 1930s. In 1955 we were married in Singapore. In 1956 she gave us a son Yik Hong 奕康. That year I was elected Associate of the Institute of Physics, London. In 1957 Lucy gave birth to a pair of twin daughters Sook Keng 淑琼 and Sook Ying 淑英. Meanwhile I carried on writing my thesis and by early 1957 I finally submitted three copies of my thesis to the university. That year I also published a joint paper with Needham and Arthur Beer, a Cambridge astronomer concerning the comet.[23] Meanwhile at the Physics Department Mr Webb had tendered his resignation and Mr Louis Hon Yung Sen 韩荣生 was appointed Acting Head.

Needham was one of my two external examiners. The other external examiner was Professor Willy Hartner, the leading authority on history of

---

[23] See Ho Ping-yü, Arthur Beer and Joseph Needham (1957), "Spiked Comets in Ancient China", *The Observatory*, 77:137–138.

Asian astronomy in Europe. It was Needham who suggested his name to Oppenheim. Oppenheim corresponded with both of them about my thesis and had probably mentioned to them that I would be due for my study leave very shortly. Needham expressed his wish to have me in Cambridge as a collaborator of his *Science and Civilisation in China* project and Willy Hartner said that I would be most welcomed to work with him in Frankfurt When I spoke to Oppenheim about my study leave, he told me about his communications with both Cambridge and Frankfurt and recommended me to go to Cambridge to gain some experience and warned me that from the tone of Hartner's letters he could sense an intention to keep me in Frankfurt. His words are, "Willy Hartner said that he would welcome you in Frankfurt. Visit him by all means, but do not let him keep you." Hartner was then Rector of a German institute of higher learning and would have resources at his disposal to offer me a job. Oppenheim was a great chess player and must have anticipated Needham's next move. But seeing that the move would be non-consequential he saw no necessity to warn me. The university granted me two years' study leave to go to Cambridge, one year being earned leave and the other being leave taken in advance, so that I was made legally bound to return to the university to serve another six years. This subtle move would foil any attempt by any outsider to take me on its payroll unless it was prepared to compensate my home university to the tune of a full year salary and allowances that I received during my leave.

The University of Malaya regarded the purpose of my leave as an opportunity for me to learn Needham's method of research and to gain some experience in Cambridge. Rayson Huang suggested that I should pay attention to Needham's unique filing system, while adding that certainly my going to Cambridge was not to learn Chinese from him. The Registrar, Hugh Lewis gave me a friendly warning saying, "Be careful! Needham is pink." To be fair to him, Needham had never tried to influence me politically, and many years later he even wrote to me saying that I was not a political animal like him. There was only a solitary incident, when he warned me about a certain visitor from America for being a suspected member of the Central Intelligence Agency. This visitor came to Needham's office in Caius College asking me for information concerning Chinese astronomical records. To me that was something quite normal and I was happy to assist him, especially when I thought that he was a friend of Needham's. After he left Needham told me that this gentleman often made trips to Soviet Russia and what was known about him was only his stated interest in astronomy, but not his employment. He concluded that a person with the freedom and resources to visit Soviet Russia would be an agent of CIA. Whether he was a CIA agent or not did not bother me at all, as it was not my concern. He carried on corresponding with me even after I left Cambridge and we became friends. I met him later several times in America. I regarded this gentleman as any other academic friend; in any case I knew of no state secret or commercial secret to divulge to anyone — except of course ancient Chinese astronomical observations, which could sometimes be regarded as top state secret in the days of imperial China. In fact, he did not even ask me about Needham. Perhaps in those days Needham could be a little sensitive; he had told me that he was blacklisted for entry to the US.[24] Of course I did not mention this incident to this American friend, but I hope that he would be amused reading this.[25]

Needham looked upon my leave as an agreement by my university to let me collaborate with him in his *Science and Civilisation in China* project and regarded my participation as a lifelong commitment. He tried to talk me out of physics by telling me that, in spite of my talent and capability, there was no hope for me to become a great physicist so long as I worked in isolation in Singapore, without ample research facilities and away from the mainstream of researchers. A few days before I was to leave Cambridge he took me out for afternoon tea at the Copper Kettle on King's Parade, across the road from Caius and facing King's College, and showed me two announcements made by Caius College — one inviting applications for the W.M. Tapp Open Research Fellowship and the other for the Berkeley Bye-Fellowship — saying that these would be the sort of opportunities I could look for. Fortunately, I could provide him a prepared answer to the effect that I was

---

[24] This was lifted in the 1970s when Needham received an honorary doctoral degree from the University of Chicago.

[25] His name is withheld only as a matter of courtesy, not that any party would be hurt by this story.

legally bound to return to my home university, thanks to the anticipation of Oppenheim. As later events would show, it turned out to be a great advantage to both of us that Needham's plan did not work out.

First Visit to Cambridge

I originally planned to be in Cambridge by September 1957. The decease of my father in June and the birth of my twin daughters, Sook Keng and Sook Ying in July the same year forced me to delay my journey to England until late November. Lucy and I had to leave our twin babies in the care of my mother in Ipoh after making arrangements to pay for the support for the three of them and a maid to help to look after our daughters. In November Lucy and I left Singapore together with our one-and-a-half year old son for Southampton by the MS *Hannover*. We arrived on the 3rd of January 1958 and took a train to Cambridge, where we were met at the railway station by Needham. He fetched us in his blue Armstrong-Siddley convertible that was previously owned by the celebrated Sir Malcolm Campbell, who broke the 300 miles per hour land speed barrier in 1935 and that of 140 miles per hour water speed in 1939. He put Lucy and my son in the front seat and used a dog chain to secure the door after closing it. My son cried, perhaps fearing that the driver was about to restrict his freedom of movement, for he had never been subjected to such treatment in my cars in Singapore. This incident seems trivial, but it served as our introduction to Dr Wang Ling 王铃. Needham explained to Lucy that the chain was installed because of Wang Ling and did not elaborate. After getting us checked in at the Garden House Hotel, he took us to 28 Owlstone Road, Newnham to meet Dr Lu Gwei-Djen 鲁桂珍. That was literally within a stone's throw away from Needham's at No. 1, Owlstone Road. Lu Gwei-Djen gave us her version of the chain in Needham's car:

> That winter Joseph took Wang Ling in his car to Oxford. Wang Ling was dozing off with one hand playing with the door hatch. Joseph suddenly saw the left-hand-side door opened and Wang Ling vanished from sight. He was worried and drove back to

look for Wang Ling. He found the latter half-buried in a pile of snow with a hand waving at him to attract attention. Wang Ling was unhurt, but Joseph still worried about what had happened. The chain was installed as a precaution against similar accident.

A few days later I met Wang Ling at Caius College. He gave me his personal version of the same incident and talked about Lu Gwei-Djen. His story about the incident is:

> Recently I nearly lost my life while travelling with Joseph in his car to Oxford. He was going there to give a talk and he was in a hurry. Suddenly the car negotiated a right curve and I was thrown out from my seat. Fortunately I landed on a pile of snow or else I would not live to tell the story. Joseph turned back to look for me and I got back in his car.

It was pointless for me to make a judgement on the two versions, but I learned the important lesson that there could be more than one likely explanation to a particular event. Wang Ling was expressing his own feelings at the time of the accident; Lu Gwei-Djen tried to explain what led to the incident that made Needham worry, resulting in the installation of the dog chain in his car that made my son nervous. One need not make a judgement unless one has to, and in any case one should guard against making a hasty conclusion. Lu Gwei-Djen and Wang Ling telling me about each other at our first acquaintance served as a warning that I ought not to participate in competitions to gain Needham's favour, which was something that I had never contemplated in any case. That was the first lesson that I learned about how to get along with Needham and his collaborators.

Wang Ling was at that time making preparations to leave for Australia to take a teaching job in Canberra at the College of Advanced Education. He told me about his frustration in looking for an employment in England. He also said that it was not through Needham but through a friend in the Faculty of Oriental Studies at the university that he finally found a place in

Australia. The next time I saw Wang Ling was in 1960 when he visited Singapore.

Wang Ling was first spotted by Needham during a visit to the Institute of History and Philology of the Academia Sinica at Lizhuang 李庄 in Sichuan province in wartime China. Wang Ling, a graduate in history, was working as a research assistant in that institute. He was sent by Dr Fu Ssu-nien 傅斯年, the Director of the institute, to go to Cambridge in 1947 to help in the *Science and Civilisation in China* project with a support of 500 pounds sterling a year from the institute. This financial arrangement did not last very long, as Fu Ssu-nien left the Chinese mainland two years later and re-established his institute in Taiwan. Needham told me that a part of his own salary from Cambridge University went to the support of Wang Ling.[26] Lu Gwei-Djen once told me that Needham had a soft spot for Wang Ling and that there were also two other Chinese friends that he was very fond of. One of them was Huang Hsing-tsung 黄兴宗, his first secretary in Chongqing during wartime China and the other was Cao Tianqin 曹天钦, the successor of Huang Hsing-tsung and once a Fellow of Caius College. Needham used to refer to Wang Ling fondly by his literary name Ching-ling 静宁, while Wang Ling would use his talent in Chinese calligraphy and poetry to write in praise of Needham. To this talent of Wang Ling I must concede. In his later years Needham often referred to the names of Lu Gwei-Djen and Wang Ling as his two collaborators.

In terms of the first person who joined Needham to work on the *Science and Civilisation in China* project, Wang Ling was the earliest collaborator. However, Needham received the inspiration to embark on the project from Lu Gwei-Djen and she gave up her work in Paris with UNESCO to join him in 1956, when Wang Ling left (not Cambridge but Needham's office at

---

[26] Wang Ling left Cambridge a couple of weeks after I first met him there in January 1958. Needham had a whole series of correspondence with the Chinese Academy of Science in Beijing (then still retaining the name Academia Sinica) about the continuation of payment to Wang Ling from China. He was trying his best not to let Wang Ling go. The archives in Cambridge and in the Chinese Academy of Science would tell further on this issue.

Caius College) and accompanied him throughout the rest of her life. In this respect she was the primary collaborator of Needham. There was a third collaborator, whom I met in Singapore a couple of years before I first went to Cambridge. He was Kenneth Robinson, who contributed the acoustic section in volume 4, part 1 of *Science and Civilisation in China*. Robinson was then with the Department of Education in Singapore as Head of the Chinese Teachers Training College at Patterson Road. He rejoined Needham in the year 1980 and helped him to mediate between collaborators as well as in the actual writing of the most important parts in volume 7 of *Science and Civilisation in China*.

Needham promised to find us a flat to rent in Cambridge before our arrival, but added that we ought to look at the place before making a decision. Provisionally he booked for us one-and-a-half room at the Garden House Hotel. However, we were kept waiting for a place to rent. After staying in that hotel for ten days the hotel bill was becoming too big for my modest budget. Lu Gwei-Djen invited us to stay temporarily at the top floor of her house at 28 Owlstone Road. After a month we moved into the second floor of number 5 Victoria Street behind Emmanuel College and just five minutes walk away from Gonville and Caius at a rent of 20 guineas a month.

When I arrived in Cambridge I thought what I would be asked to do would at least have some remote connection with physics. However, Needham said he had already sent his manuscripts on the mathematics and astronomy sections to the press, and that all the required material for the physics, engineering and shipbuilding sections were already assembled for him to work on. He then asked me to go through the *Daozang* 道藏, a collection of more than 1,100 volumes of Daoist text, to dig up information for the future alchemy section. Having studied chemistry for only one year at Raffles College, I had little confidence to do any research on alchemy. One day I met Professor J.R. Partington at Gonville and Caius College in Room K1, Needham's office then. Partington was formerly the Professor of Chemistry at Queen's College, London University and had written several chemistry textbooks for university students. He later became an authority in the history of chemistry and lived in Cambridge after his retirement. Needham often consulted him on matters concerning alchemy and the history of chemistry. I first came across Partington's name in a chemistry textbook that I saw the first day I went to the library in Raffles College, but when I went to the library the next day to borrow the book, it had already disappeared from the shelf. After telling him this little story I turned to him for advice on my lack of self-confidence to work on the history of alchemy, as I had already forgotten what little I learned about chemistry in my first year in college. He replied by setting me a test, asking me whether I knew what H_2O was. When I answered in the affirmative and told him that I still remembered a few more chemistry formulas, he smiled and said that I had already passed my test with flying colours, because it was not necessary to know any chemistry formula to do research in alchemy.

I received encouragement to do research on alchemy from the first big name in chemistry I came across in my student days in Singapore. I also had an encounter with an even bigger name in physics, but learned another kind of lesson from him. In my Honours year I learned about the neutron and its discoverer, Sir James Chadwick, a Nobel Laureate. Professor Alexander asked a former fellow research student in the Cavendish Laboratory to be my external examiner. Hearing that my would-be external examiner was a student of Chadwick, I read up Chadwick's publications on the neutron in the *Proceedings of the Royal Society* and other journals. Thus I knew much more about Chadwick's name and his work than about Partington. Shortly after my arrival in Cambridge, Needham invited me to dine in Gonville and Caius College. I sat at high table near the corner on the right hand side of an elderly-looking gentleman wearing a cap that gave me the impression of being the chaplain of the college. After spending some time talking to the person on his left he turned round to look at me and asked what I was doing. I answered saying that I was a lecturer in physics in Singapore. He took no notice and remained silent. When Needham pointed out to me that this gentleman was Sir James Chadwick, the Master of the college, I quickly told Chadwick the name of my physics professor, who was trained in the Cavendish Laboratory. Again he remained silent. Thinking that I might be able to refresh his memory, I gave him the name of my external examiner, a former student of his. Then he responded by criticising both the professors whom I had great respect for, making me feel smaller and smaller. Needham came to my rescue, introducing me to him as his new collaborator who had discovered the Chinese parhelia. Chadwick blamed my external examiner for missing the positron, but when I mentioned that Anderson first discovered it, he said that he himself was the first discoverer. Perhaps he did know about the positron before C.D. Anderson, but the latter published his results earlier. My encounter with Chadwick did not diminish my respect for my physics professor and my external examiner, but it made me aware of peerage ranking in the academic world. On the positive side, after surviving this baptism of fire I found myself able to dine at ease at any college high table, and I learned to make my guests dine at ease when I later became Master of Robert Black College in Hong Kong. After my experience with Chadwick I had innumerable congenial occasions dining at Caius College. I had dining rights at the College as a Member of the Senior Common Room. Figure 3

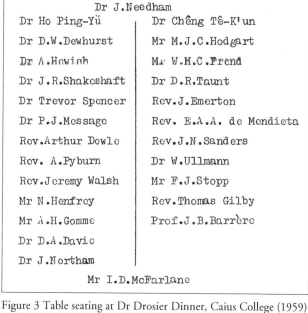

Figure 3 Table seating at Dr Drosier Dinner, Caius College (1959)

(Note that Needham persisted in writing Chinese names, such as those of Cheng Te-k'un and mine, in the system he preferred, but he was not consistent in this matter. He did not do it to Lu Gwei-Djen, nor did he venture to standardise names other than Chinese to any particular system.)

shows the arrangements on the table presided by Needham at the Dr Drosier Commemoration Dinner on 2 May 1959. Opposite me was Dr Cheng Tu-k'un and to my right were Dr Dewhirst and Dr A. Hewish, both astrophysicists. Dr Hewish later won the Nobel Prize in 1974 for his discovery of quasars.

Encouraged by Partington, I spent about half the time every day in the University Library combing through Daoist texts searching for references to Chinese alchemy and making notes and translations. Fortunately only a very small fraction of those texts were found relevant to my studies. However, part of the material I collected enabled Needham and me to write and publish four articles that would form parts of the future alchemy section of *Science and Civilisation in China*.[27] I also published a joint paper with Needham on Chinese meteorology from material taken out from my thesis and another joint paper with Needham, Edwin Pulleyblank, the Professor of Chinese in Cambridge University, two Cambridge astronomers and Lu Gwei-Djen on the 8th-century measurement of the meridian line performed by the Chinese monk-astronomer, Yixing.[28] Then there was also a joint paper on the Chinese aurora with Dr Justin Schove.[29] However, among my publications I derived the greatest satisfaction from my catalogue of Chinese comets and novae that appeared in the British journal *Vistas in Astronomy*.[30]

---

[27] See Ho, P.Y. and Needham, J. (1959), "The Laboratory Equipment of the Early Mediaeval Chinese Alchemists", *Ambix*, 7:57–115; Ts'ao, T.C., Ho, P.Y. and Needham, J. (1959), "An Early Mediaeval Chinese Text on Aqueous Solutions", *Ambix*, 7:122–158; Ho, P.Y. and Needham, J. (1959), "Elixir Poisoning in Mediaeval China", *Janus*, 48:221–251; Ho, P.Y. and Needham, J. (1959), "The Theories of Categories of the Mediaeval Chinese Alchemists", *Journal of the Warburg and Courtald Institute*, 22:173–210.

[28] See Ho, P.Y. and Needham, J. (1959), "Ancient Chinese Observations of Haloes and Parhelia", *Weather*, 14:124–134 and Beer, A., Ho, P.Y., Lu, G.-D., Needham, J., Pulleyblank, E.-G. and Thompson, G. (1961), "An 8th-Century Meridian Line", *Vistas in Astronomy*, 4:3–28.

[29] See Schove, J. and Ho, P.Y. (1959), "Chinese Aurorae I: A.D.1048–1070", *Journal of the British Astronomical Association*, 69:295–304.

[30] See Ho, P.Y. (1962), "Ancient and Mediaeval Observations of Comets and Novae in Chinese Sources", *Vistas in Astronomy*, 5:127–225.

At that time I thought that such a publication on astronomy would only bring me a little closer back to physics, but to my delight it turned out to be useful to both astronomers and astrophysicists. I tried to keep up a little with physics and attended the early morning 9 o'clock lectures on solid-state physics given by Professor Nevil Mott, who later won the Nobel Prize in 1977. Needham was completely unaware of this, since he turned up in his office normally around 11 o'clock. I also helped my university to get Dr David Shoenberg, Reader in Physics at Cambridge University, to serve as an external examiner for physics in Singapore. When I left Cambridge towards the end of my study leave I thought that my collaboration with Needham had ended, and that I would return to Singapore to teach physics. I handed over to Needham the notes and translations I made on the Daoist alchemical texts thinking that they would serve a better purpose in his library since I would be working on something less remote from physics than alchemy.[31]

Needham was away from Cambridge for several months, both in 1958 and in 1959. He spent the first period in Sri Lanka, then known as Ceylon. Lu Gwei-Djen accompanied him in the second period of their China tour. They stopped by Singapore on the way home and visited the libraries of the two universities in Singapore as well as the Botanical Garden. In Singapore they were guests of Oppenheim and Needham reported to him about my work in Cambridge. They also went to see a mutual friend Ivy Soh, an English language specialist trained in England. Needham and Gwei-Djen brought back many Chinese books, which they stamped with a seal that read *zou ma kan hua* 走马看花 — looking at flowers on a running horse — to mark the occasion.

I was left to work entirely on my own. Needham left word that should I need help in chemistry I could ask Dorothy Needham to take me to the Faculty of Biochemistry. I did get her help to take me to the library of that Faculty to look up some points in inorganic chemistry from Mellor's huge compendium on that subject. It occurred to me that only Dorothy Needham maintained good contact with her colleagues and friends at the Faculty of

---

[31] I am glad that they had served a useful purpose. See, for example, Needham, J. and Lu G.-D. (1983), *Science and Civilisation in China*, 5.5: xxxiii and 455.

Biochemistry. I had not heard Lu Gwei-Djen mentioning the names of those she worked with before in the Faculty of Biochemistry other than those of Joseph and Dorothy Needham.

Working on my own had its advantages. I came to know several members of the Faculty of Oriental Studies while working in the Chinese section of the Cambridge University Library. Dr Greta Scott was in charge of this section where only readers granted approval could gain admittance. Only a few readers were there at a time and I came to know all of them. Sinologists formed only a small circle so they desired to know one another. During teatime we joined one another in the tea room and sometimes Greta Scott would also join us. That was how I first met Van der Loon, who was advising the Chinese section at that time and who later became Professor of Chinese at Oxford. It was also there that I first met Dr Denis Twitchett, who later became Professor of Chinese at the School of Oriental and African Studies, where he had come from earlier, before returning to Cambridge as Professor of Chinese until he moved finally to Princeton in the early 1980s. I also knew Dr H.C. Chang 张心苍 of the Faculty of Oriental Studies, being former colleagues in the University of Malaya, where he was an English lecturer. It was a coincidence that when I returned to Singapore later, the university allocated me house number 8 as my living quarters. This was the same house occupied by H.C. Chang before he left Singapore.

I met, however, the two most senior Sinologists of the Faculty at Needham's office, Room K1 in Caius College. They were Professor Edwin Pulleyblank and Dr Cheng Te-k'un 郑德坤, Reader in Chinese Archaeology. Pulleyblank was working together with Needham and a small group including myself on the 8th-century Chinese meridian line measurement. Cheng Te-k'un was working on the series *Archaeology in China*, which was hailed at that time as one of the two major contributions by Cambridge to Chinese Studies, the other being Needham's *Science and Civilisation in China*. Needham had much respect for Cheng Te-k'un, telling me that the latter was a scholar and a gentleman. He invited both of us to dine at Caius College. My family and I visited Cheng Te-k'un and Mrs Cheng at their home at 166 Chesterton Road in Cambridge. Our friendship later carried over first to Kuala Lumpur and then to Hong Kong.

Needham introduced me to his filing system soon after I joined him at room K1 in Gonville and Caius College. He also had the use of room K2 next door, which was occupied by Lu Gwei-Djen. He let me use a small desk next to his table in room K1. His working library was housed in both K1 and K2. The latter held mainly medical and pharmacological works as Lu Gwei-Djen was responsible for the medical volume in the *Science and Civilisation in China* series. Needham would make a card for every new book title, personal name and geographical name he came across while working on the project. He translated these into English and transliterated every book title using his own modified Wade-Giles system, and attempted to find its author and the year the book was written. Similarly he transliterated every personal name and tried to find the period the person was in. He did likewise to geographical names and tried to find their modern locations. He also prepared a duplicate set on coloured cards, so that he could let his cards be taken out of his office for preparing the index. He called the duplicates "ghost cards". He also had a set of cards for technical terms. Whenever he came across a name or a term that sounded a little unfamiliar, or when he wished to refer to something, such as the dates, he would first look up his cards.

It goes without saying that I made full use of this system while I was working with Needham in Cambridge, but I did not make any attempt to duplicate it as it would not be practical for a lecturer in physics to do so in my home university. I would rather adapt it to suit my own circumstances. As far as I know, only Wang Ling was concerned with this exercise, and for many years he had a research assistant to help him in his work. That was a remarkable system at the time Needham invented it, but a knife does not cut both ways. At the beginning its advantages were many. It was originally planned to serve the writing of only seven volumes, but Needham eventually saw over twenty sub-volumes. The number of cards increased until they filled up scores and scores of boxes and took up more and more time to prepare, to search and to read. Being no longer easily portable, Needham could not work on his project if separated from his cards. For the sake of uniformity, every collaborator in the *Science and Civilisation in China* series would have to refer to the cards. This became a problem if one did not live in Cambridge. In our current computer age we can of course find a much better solution. I remember that it took me three months to calculate my data on an electric calculator to obtain the result for my first publication in a British journal. I would imagine that three hours would suffice to put the same data on an electronic spreadsheet to obtain the result. When Needham visited Hong Kong in 1981, his old friends, Mr P.L. Lam 林炳良 and Mrs Mary Lam took him to a computer shop and bought him a present in the form of an IBM XT personal computer, hoping that he would put it to good use. Alas, it came far too late for a person over 80 years of age to learn a new technique and to replace his thousands and thousands of cards with electronic files. Needham brought the computer home, but did not use it personally.

Needham made notes of every relevant book he read on pieces of paper and kept them in files. He also kept reprints in files, sometimes with marginal notes or comments added. He employed his own system to arrange his books and files according to the sequence of the sections in the *Science and Civilisation in China* series. He remembered the exact positions of his books and files. I was very careful to put everything back to its original position. Occasionally Lu Gwei-Djen would tell me, "Joseph is in a bad mood today". Needham tried to explain to me that he was not angry with anybody but was upset that his memory was failing him.

During the absence of Needham from Cambridge I sent him drafts of translations and notes I made on the Daoist texts that I read in the Cambridge University Library. On his return to Cambridge Needham started to work with me on the preliminary papers for the alchemy section of volume 5. He and I sat together from the afternoon till dinnertime, when he would dine in College and I would return to 5 Victoria Street to join my family for the evening meal. After dinner we would meet again at Caius College and worked until about 11pm. This went on for about six weeks. Naturally I had to adjust myself to this schedule, but Lu Gwei-Djen complained to me that Needham came to see her much later than his routine after-dinner visit from the College.

When we worked together Needham would bring out a pair of scissors, a bottle of paste, and the translations and notes I sent him which he would have already read. He would jot down the points he wished to write about

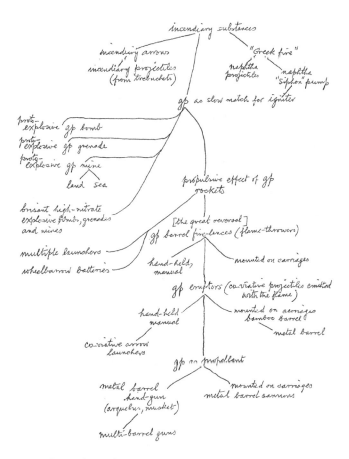

Figure 4 Joseph's railway lines in his own handwriting

on a piece of paper and connect them in sequence by lines, which Lu Gwei-Djen used to tease Needham about, calling them "Joseph's railway lines", because of Needham's fondness of the railways. (See Figure 4.) By applying the cut-and-paste method, sometimes changing the style of writing in my original work, and incorporating his own ideas and new findings, Needham and I managed to write four papers for publication. Needham called his process of joining facts together in writing the "weaving method". I can appreciate the cut-and-paste facilities of the modern word processor all the more when I think of the days working together with him.

I managed to find some time away from work during my stay in Cambridge. I bought an export version of the Morris Minor 1000. With it I took my family to either the zoo at Regent's Park, London, or that in Whipsnade, or to the seaside during the weekend, sometimes bringing along a friend. Lu Gwei-Djen joined us on a couple of occasions. Abdullah bin Ayub, an old friend from Perak and senior civil servant with the Malaysian government, had also been our guest. On another occasion I took my family by car to Brinsford Lodge near Wolverhampton to fetch my sister Lay Kum to Cambridge. Lay Kum was at Brinsford Lodge undergoing a teachers' training course. We also visited the English lakes as well as the lochs of Scotland. We went as far north as Stirling.

While in Cambridge Lucy and I kept in contact with students from Singapore and the Federation of Malaya studying in Cambridge. They formed a Cambridge University Malayan Students Association. The President of the association at that time was Chua Seong Siew and the Honorary Secretary was Lim Pin 林彬, both of whom were medical students. Chua Seong Siew later practised as a heart specialist in Kuala Lumpur and Lim Pin was once the Vice-Chancellor of the National University of Singapore before resuming his former post of Professor of Medicine. Lim Pin asked me to give a talk on higher education in Singapore on behalf of the association. The talk took place in his room at Queens' College. I did not need to count to know the number of my audience — besides the President and the Honorary Secretary I saw only one other Cambridge undergraduate in the room. Nevertheless I delivered my talk. Question time followed and there came one from the undergraduate asking me to compare the standard of the University of Malaya with that of Cambridge. I replied that my answer would be equally unfair as the question asked, saying that the University of Malaya was then only 10 years old while Cambridge was some seven centuries old, but if a comparison of standard was to be made, I would say that the former after 10 years had achieved a higher standard than the latter after five centuries.

I met with another adventure in lecturing. This time was at Kent, just outside London. Dr Justin Schove, the Principal of St. David's College at Kent, used to go to Needham's office at Caius to ask for information on Chinese astronomical records. Needham understandably had little time to help him to search for information in the Chinese texts and to do the necessary translations for him. He found me more obliging and with the information I supplied, we published several joint papers on the Chinese aurora. Through him I was elected member of the British Astronomical Association in February 1959. One day he came to Caius College and invited me to give a talk at his College. I thought that I could talk on the same topic that I gave at Queens' College so as not to waste it on only three listeners. I brought my notes along and drove my new car, a Morris Minor 1000, all the way from Cambridge to Kent. As I entered the college grounds I saw only groups of young children. I soon realised that the college was a kindergarten for pre-school children. Justin Schove introduced me to the children as Dr Ho from Singapore. I had to think quickly on what to say to them. I took out my lecture notes, but used the paper on which my intended lecture was written for a demonstration in Chinese paper folding, popularly known by the Japanese term *origami*. The children seemed to enjoy that session and Justin Schove said that he would like me to speak to the children again. I politely declined his invitation saying that while I enjoyed speaking to the children, time would not permit me to indulge in such luxury.

In the summer of 1959 Professor Kiyosi Yabuuti paid a visit to Caius College prior to his attendance of the 7th International Congress of History of Science, which was to be held in Barcelona and Madrid in September. This was the first time Needham met him and that also applied to me. Dr Shigeru Nakayama 中山茂 also visited Caius College. He told me that although that was the first time we met, he knew about me earlier from the draft of my translation of the *Jinshu* that Needham showed him when he was in Cambridge in 1957. Nakayama became the first Japanese academic associate I had a long-lasting friendship with.

Dorothy Needham and Lu Gwei-Djen travelled by car to Barcelona. My wife and I, together with our son, travelled in our own car, going right across France to Spain to attend the International Congress of History of Science. There we met Kiyosi Yabuuti and Shigeru Nakayama again. It was also there that I first met Willy Hartner, who invited me to visit Frankfurt during my next tour to Europe. We did not go to Madrid, but travelled across the south of France to Genoa in Italy to board the MS *Asia* for

Singapore. Needham went to Madrid together with Dorothy and Lu Gwei-Djen, but caught up with us in Cannes to bid us farewell.

Readership in the University of Singapore

I returned to Singapore in December 1959 when the university term was about to end. I took advantage of the term break to take Lucy and our son to Hong Kong on a short holiday travelling onboard the MS *Victoria*. On my return to Singapore I received a letter from Needham saying that he had received a generous donation from Dato' Lee Kong Chian 李光前 of Singapore. That probably inspired him to set up a "Friends of the Project Committee" led by Dr Victor Purcell to appeal for funds to help him in his work. He asked me to send him a photograph taken at a photo studio so that Cambridge University Press could include it together with his own and those of his three other collaborators in a prospectus about the *Science and Civilisation in China* series. When I first joined Needham in Cambridge the series was only planned for seven physical volumes. Needham once expressed to me his desire to see the whole series completed before he reached the age of 70 in view of the fact that one of his parents died before the age of 60 and the other before 70, and that his physical build was unlike the slim figure of Partington which would enable him to live to a ripe old age. After the publication of volume three of the series, however, Peter Burbidge of the Cambridge University Press, who was responsible for publishing the series, told Needhm that his book had grown too "fat" and needed to be trimmed down in size. Needham was at first taken aback, but became delighted when Peter Burbidge suggested that he could bring out as many sub-volumes as he needed. The series was selling well; it would do the press no harm to have more from Needham. Some of the subsequent volumes, such as volume 4 part 2 almost reached the size of volume 3. I heard no complaint from Peter Burbidge. Needham seemed to have forgotten his fear of not seeing the whole series through; perhaps he began to have more ideas on looking for more collaborators to let them finish the job. In 1960 the Cambridge University Press published the prospectus for the series which showed volume 4 in three parts and volumes 5 and 6 in parts yet to be determined. The last page carried a photograph of Needham at work together with his four early collaborators.

By that time the University of Malaya was divided into two divisions, one in Singapore and one in Kuala Lumpur. They shared a common Vice-Chancellor and each division had its own Principal. Oppenheim became the Vice-Chancellor and was stationed in the Kuala Lumpur Division, while Professor S. Sandosham became the Principal of the Singapore Division. In 1960 I was appointed Reader in History of Science within the Physics Department, which had a new Head in Professor K.M. Gatha. Readers were "rarities in Malaya in those days", to borrow a phrase used by Rayson Huang, although it is applied here to a slightly lower level in the academic ladder. Besides me, there was only one other Reader in the university. He was none other than Dr Toh Chin Chye 杜进才, who once shared the same household with me together with four other young bachelor members of the university teaching staff. We were living in three university quarters, jointly employing one domestic servant and had our meals together. He was Reader in Physiology, but subsequently became the Deputy Prime Minister of Singapore and then the Vice-Chancellor of the University of Singapore.

The term 'Reader' is not often understood outside Britain. Professor Jao Tsung-I 饶宗颐, formerly Professor of Chinese at the University of Singapore, told me that when he was Reader at the University of Hong Kong, he went to attend to some matter at a government office. When he told the clerk in the office that he was a Reader the latter asked him to stop joking, because he himself was also a reader of magazines and newspapers. The duties of Readers vary from university to university. It is interesting to compare a Readership in Cambridge, taking Needham as a case study, with a Readership in Singapore, taking my own case, which did not necessarily apply to that of Toh Chin Chye. The Cambridge system gave the Reader almost complete freedom to do what he wished to do. As long as he attended to his normal teaching of biochemistry undergraduates he had no further obligations to participate in other duties that his colleagues had to perform and he was free to pursue research even outside biochemistry. Needham took full advantage of this system, but everything has a price to pay. We cannot expect a person to be popular among colleagues who have to share his workload, nor can we expect any department head to allocate funds or apply for grants to support research projects outside the domain of his department, and when the position of department head fell vacant Needham had not been seen inside the biochemistry laboratories for more than 20 years. Not every Reader took full advantage of his privileges as Needham did, but without these privileges it would not have been possible for him to embark on his *Science and Civilisation in China* project. As Mencius said, it was a choice between fish and the paw of a bear; when he could not have both he would take the paw of the bear and leave the fish alone. Through the eyes of East Asia and the history of science, like the ancient Chinese sage, Needham had made the wise move of taking "the paw of the bear". However, the fame attained by him led some in East Asia to forget that he had already abandoned "the fish".[32]

In contrast to Needham's Readership, my post did not relieve me of any part of the normal duties of a member of the teaching staff in the Physics Department. On the contrary, I was given the duty of running a sub-department and organising a one-year course on the history and philosophy of science as an optional second year subject within the Science Faculty. The Faculty of Science also elected me Chairman of the Science Faculty Public Lectures Committee, while the university Senate made me its representative on the University Library Committee to offer advice on its Chinese collection. I cannot remember how I managed to find all that energy. I was then in my mid-thirties. I was involved in a number of government boards and committees, mainly in the area of education, for example, the Education Advisory Board, Universities Scholarships and Bursaries Board, and the Textbooks and Syllabuses Sub-Committees. I even served as advisor to the Examination Syndicate in Kuala Lumpur. With all these activities I still found time to enjoy playing tennis or table-tennis with my friends and to have an occasional game of chess, representing either the university or its graduates in tournaments. I was Chairman of the Academic Staff Association Tennis Club. The Club arranged social matches to meet players from the Singapore Police, the British Navy and Air Force, and schools. I was the

---

[32] I have often heard people in East Asia asking why Cambridge University did not offer the Chair of Biochemistry to Needham.

weakest player in my team and was invariably made to play the third single. This turned out to my advantage, because I seldom lost my game to my counterpart on the opposite side of the court. I knew my own limitations in tennis. My physical build did not enable me to develop a powerful service, and hence I tried to concentrate on placing and consistency that would enable me to give a lesser player a good game. After all tennis was only a form of exercise for me, if not a recreational cum social meeting with friends. In fact tennis was one of the things I missed since I left Singapore in 1964. Without the right group of friends playing together, I found little incentive and opportunity to play the game. As for table-tennis, there was a table-tennis room just next door to my university quarters. I could easily walk across and join in when I heard activities generated from that direction.

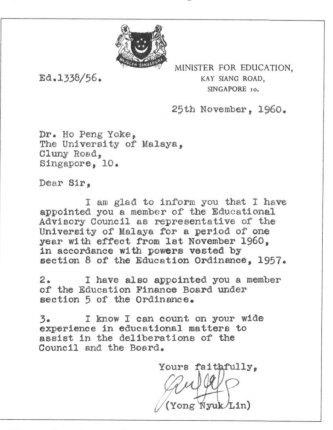

Ed.1338/56.

MINISTER FOR EDUCATION,
KAY SIANG ROAD,
SINGAPORE 10.

25th November, 1960.

Dr. Ho Peng Yoke,
The University of Malaya,
Cluny Road,
Singapore, 10.

Dear Sir,

I am glad to inform you that I have appointed you a member of the Educational Advisory Council as representative of the University of Malaya for a period of one year with effect from 1st November 1960, in accordance with powers vested by section 8 of the Education Ordinance, 1957.

2.   I have also appointed you a member of the Education Finance Board under section 5 of the Ordinance.

3.   I know I can count on your wide experience in educational matters to assist in the deliberations of the Council and the Board.

Yours faithfully,

(Yong Nyuk Lin)

Figure 5 Letter of appointment from Mr Yong Nyuk Lin, Minister of Education

The most important chess event for me then was the annual Town versus Gown match where the whole city took on the staff and students of the university over some 30 boards. The captain for the university side was Dr Lim Kok Ann 林国安, Professor of Bacteriology and a grandson of Dr Lim Boon Keng. As a chess enthusiast, he went to serve the world chess organisation FIDE after his retirement from the university. As captain he had to play the first board to take on the strongest opponent. Oppenheim often played either the second or the third board, depending on the strength of the top player among the students. I was placed to play between the 6th and the 10th board. My first game was a disaster. My opponent was a new arrival to Singapore whom I later discovered to have been the Irish Correspondent Chess Champion. He was playing white and started off with a king gambit that I had never come across before. After accepting his gambit I found myself desperately on the defensive and the result was a forgone conclusion. After the match I quickly bought a few books on chess openings to improve my game. After my first lesson I had not lost another game in the Town versus Gown series, although the Town team invariably came out on top in every contest. I won most of my games and drew only in a few. I also joined the graduates, mainly those from the Faculty of Medicine, to participate in a city open chess tournament, playing the first board. Our team emerged as Division Two Champion. Again I did not have much opportunity to have a game of chess after leaving Singapore in 1964. I did have a few games of chess at the University House of the Australian National University in Canberra when I was there as the 1972 Asian Fellow of that University. My last game of chess was in 1975 when I played the first board as white against the students' champion of the School of Science at a Staff-Students match of Griffith University in Australia. I won by using the Bird's Opening that my opponent was unfamiliar with.

Every year the Head of State of Singapore invited those who had rendered service to the state to an outdoor dinner party at his official residence, the *Istana* (palace), where guests were free to move about and select their food and drinks from various stalls that served local and Western food. Among them were those that served *satay* (skewered barbecued meat, mainly lamb, beef and chicken) prepared by the traditional Malay method. My wife and

I used to look for these stalls together with some of the invited guests. Mr and Mrs Lee Kuan Yew 李光耀 were also present on such occasions to greet the guests. In 2001 I sat in a restaurant at the Holiday Inn Parkview Hotel in Singapore and had a good view of the trees hiding the *Istana* across Clemenceau Avenue. It brought back memories of some 40 years ago when I had on several occasions driven into the grounds behind those trees with Lucy to attend what was our most important annual social function of the year.

We moved into house number 8, on our return from England. It was a large and solidly built colonial house. We employed a domestic servant and a washing woman to help in the house. When we returned from Hong Kong we went back to Ipoh to fetch our two daughters Sook Keng and Sook Ying. Our family lived together again. We sent Yik Hong and his two sisters to the kindergarten at Dean School. Their cousin Vivien Lee came all the way from Seaview Park, Katong, to join them at Dean School. It became a routine for me to take my family to Katong every weekend to visit Vivien's family and my father's cousin Qixun, who was in Singapore with his wife and his aged mother. Coinciding with my promotion to a Readership my wife gave birth to our third daughter Sook Kee 淑姬.

On the financial side my Readership carried a higher salary than my former Lectureship, and at the same time my brother Ho Thong, who graduated from medical school, took over part of my burden to contribute to support my mother. Shortly afterwards I took my whole family to Penang to attend the wedding ceremony of Ho Thong and Margaret Goh. I was able to have a little more savings and started to buy my first lots of shares. I first heard about buying company stocks and shares from Lu Gwei-Djen in Cambridge. She told me that she deliberately kept away from the shares of companies that manufactured weapons of war. I had no need to take heed of her advice, since I did not know of any public-listed company that manufactured weapons of war in that part of the world. It was Lucy's brother-in-law and my former school classmate, Lee Kam Wai who introduced me to his sharebroker and recommended me to buy my first lot of shares. I did not waste much time studying the share market, since I had reliable knowledgeable friends to advise me. I began to find myself on a sound financial footing and was even seriously contemplating investing in a house of our own in 1963, when a number of university colleagues joined together to build their own houses in a housing estate in Eng Neo Avenue, Singapore. It was my appointment to a new chair in Kuala Lumpur that altered this plan. All in all, despite my heavy workload, 1960 was a very memorable year to us. I had no other ambition then but to carry on working in Singapore.

With my Ph.D. thesis behind me I now had the freedom to get more in touch with the community in Singapore. That provided me an opportunity to practise writing in Chinese. Through the introduction by one of my first year medical students, Lien Liang-ssu 连亮思, I met his father Mr Lien Shi-sheng 连士升, Chief Editor of the *Nanyang Siangpau* 南洋商报. Mr Lien asked me to write an article on Needham for the New Year issue of his newspaper. I next met Mr Liao Sung-yang 廖颂扬, Manager of the *Sin Chew Jih Poh* 星洲日报 at a committee meeting in the university. Liao Sung-yang said that he saw my article in the *Nanyang Siangpau* and asked me to write one for his newspaper as well. This was a request which I was not in a position to turn down. In this way, for about the next 10 years, I had to write two articles in Chinese every year. I treated all these as opportunities for me to practise writing in Chinese, without thinking that my effort would serve to introduce me to the Chinese-educated communities both in Singapore and in the Federation of Malaya. My first three or four articles were written in literary Chinese. I thought that, being only a scientist, it was not my business either to follow or to set a trend in the writing of Chinese; I could therefore choose to write in whatever style that pleased me. In 1963 I visited my younger sister's husband, Ong Ah Chan 王亚灿, in Kuala Lumpur. Ah Chan was a successful motor vehicle spare parts dealer, although he had not received very high education in either Chinese or English. He told me that he saw a very good piece of article written by me in the newspaper. I asked him in what way did he find it good. He answered innocently, saying that my article was written in such a profound style that he could not comprehend it. That gave me a shock and opened my eyes. The newspaper is a mass medium for communication. An article that does not communicate well should not find a place in a newspaper. From then onwards I started writing in spoken Chinese.

In 1961 a colleague from the Mathematics Department, Oon Lay Yong 温丽蓉 (later became Lam Lay Yong 蓝丽蓉 after her marriage) asked me to be her supervisor for her doctoral thesis on the 13th-century mathematician Yang Hui 杨挥. One day she sought my advice about getting access to a handwritten copy of Yang Hui's work on magic squares which was owned by Li Yan 李俨 in Beijing. She thought of asking her maternal grandfather, Mr Tan Kah Kee 陈嘉庚 in Beijing to get a microfilm copy of the book from Li Yan, but hesitated to write because of Mr Tan's old age. Mr Tan Kah Kee was then the most respected and influential overseas Chinese in China. I told Lay Yong that I might be able to help avoid troubling him by writing to Needham. At the request of Needham, Li Yan sent two copies of microfilm of Yang Hui's work on magic squares to Cambridge; Needham directed one copy to Singapore. Lam Lay Yong later became Professor of Mathematics at the National University of Singapore and after her retirement she won the Kenneth O. May medal, the highest award of the International Commission on the History of Mathematics, in 2001 in Mexico City, for her contributions to the study of Chinese mathematics.

In 1962 the Singapore Division of the University of Malaya became the University of Singapore with Dr. B.R. Sreenivasan as Vice-Chancellor, while the Kuala Lumpur Division retained the name University of Malaya and had Oppenheim, by that time known as Professor Tan Sri Sir Alexander Oppenheim, as Vice-Chancellor. The same year I made my first visit to the US to attend the 10th International Congress of History of Science held in Ithaca, N.Y. and Philadelphia. There I presented a paper on Vietnamese astronomical records. Professor Edward Schafer asked me to expand my presentation for publication in the *Journal of the American Oriental Society*, of which he was the editor.[33]

Before I left for the US I received a letter from Needham asking me to have a word with Dr Cammann Schuyler of the University of Pennsylvania, who had made some criticisms on the mathematics section of *Science and*

---

[33] See Ho, P.Y. (1964), "Natural Phenomena Recorded in the *Dai-Việt Sử-Ký Toàn-Thư*, an Early Annamese Historical Source", *Journal of the American Oriental Society* 84:127–149.

Civilisation in China volume 3 in a book review. Schuyler pointed out a mistake in one of the Chinese magic squares and the wrong name "Li Nien" that was given to the Chinese historian of mathematics, Li Yan. Although I was not involved at all in the mathematics section, I met Schuyler in Philadelphia and explained to him that the same magic square he and Needham talked about contained a misprint itself, and as for the name of the Chinese historian of mathematics, Li Yan was correct, and that Needham had probably listened to a native from a certain part of Jiangsu province reading the name in his own dialect, and after having the name in that form it would be cumbersome to make a change. Schuyler accepted my explanations and showed me round the museum in the University of Pennsylvania. I reported to Needham about my meeting with Schuyler, but that gave him the wrong impression that Li Yan was a native of Jiangsu province, although she actually came from the Fujian province. Needham first met Li Yan in 1956 at the 8th International Congress of History of Science held in Florence, but he told me that he had never met Li Nien, although he had corresponded with the person in 1961 at my request concerning the microfilm on Yang Hui's work on magic squares. A great man is also entitled to possess some idiosyncrasies of his own.

On my return from America I found a research student from Harvard University waiting for me. Nathan Sivin, who was introduced to me by Shigeru Nakayama, was sent to Singapore with a Fellowship from his university to spend the better part of the year with me to work on his Ph.D. thesis on Chinese alchemy. Sivin is now the most senior professor and authority in the history of Chinese science in the US. A third Ph.D. candidate holding a research fellowship from a Christian church approached me to act as his supervisor, but the topic he suggested had nothing to do with the history of Chinese science, but with science and religion instead. I told him that I was not competent to help him to reconcile two parts of his topic that required different mental approaches; one was scepticism and the other faith. Eventually he changed his mind and did a Ph.D. degree in chemistry with someone else instead.

I used to relieve Needham of part of his burden by going back to the grounds he had previously covered. He would always like to look ahead while working on his *Science and Civilisation in China* series and was reluctant to go back to the volumes that he had previously written. His preference was to tread on virgin grounds, so he said. He used to pass queries and consultations on matters concerning Chinese astronomy to me for handling. Figure 6 shows a typical example. Dr Dewhurst was an astrophysicist, sitting between his colleague Hewish and me at the Dr Droiser Commemoration Dinner at Caius College in 1959 as shown in Figure 3.

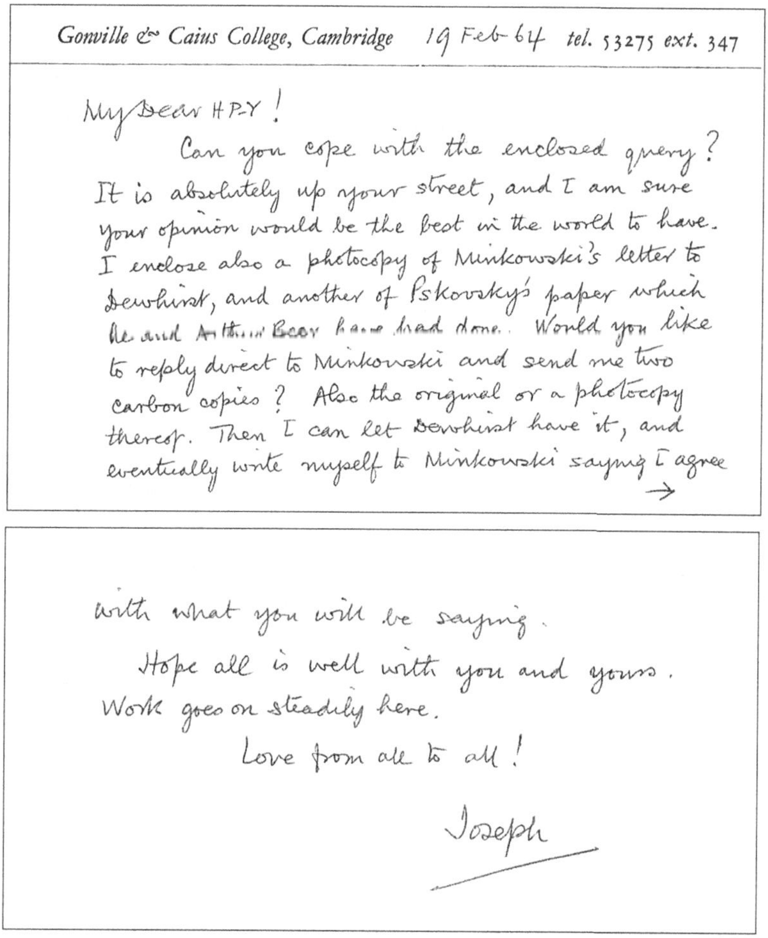

Gonville & Caius College, Cambridge   19 Feb-64   tel. 53275 ext. 347

My Dear H.P-Y !
    Can you cope with the enclosed query? It is absolutely up your street, and I am sure your opinion would be the best in the world to have. I enclose also a photocopy of Minkowski's letter to Dewhurst, and another of Pskovsky's paper which he and Arthur Beer have had done. Would you like to reply direct to Minkowski and send me two carbon copies? Also the original or a photocopy thereof. Then I can let Dewhurst have it, and eventually write myself to Minkowski saying I agree →

with what you will be saying.
    Hope all is well with you and yours.
Work goes on steadily here.
        Love from all to all !

                Joseph

Figure 6 Postcard from Needham in his own handwriting

The first Chancellor of the University of Singapore was Dr Dato' Lee Kong Chian 李光前. Late in the 1950s Wu Lien-teh had told him about Needham's project on the history of Chinese science and asked him to render a helping hand to Cambridge. In November 1959 Needham received from Dr Dato' Lee a "munificent gift", to use his own words. It was likely that Wu Lien-teh had mentioned about my collaboration with Needham. This would explain why my Chancellor showed much interest in my research and became quite friendly with me. He invited me many times to have lunch with him at the old OCBC Building in Chulia Street. I usually went together with Professor Lim Tay Boh 林溪茂, who became the Vice-Chancellor after I left Singapore in 1964. On one occasion my Chancellor introduced me to Tan Sri Tan Chin Tuan 陈振传, his trusted friend and assistant who later succeeded him as Chairman in the Board of Directors of OCBC Bank, of which the Lee family has always been the majority shareholder. I taught his eldest daughter, Keng Lian, physics when she was a first year medical student in Singapore.[34] Lam Lay Yong was a niece of Lee Kong Chian and a close friend of another daughter of Tan Chin Tuan. This paved the way to Tan Chin Tuan's donation to the Needham Research Institute in 1984. While I was in Cambridge during the 1990s I had several opportunities to talk to British academics about my Chancellor, not about his invitations to lunch, but on his personal interest in learning and in the academic activities of the university. I told them that I once invited Professor Sir Harrie Massey, FRS, to give a public lecture on space exploration at the University of Singapore. My Chancellor heard about this, but found that he could not be present at the lecture because he had to chair a business meeting. He spoke to me over the telephone to ask me to make a tape-recording of the lecture for him. One of my British colleagues remarked how much he wished to have a Chancellor who took so much personal interest in the academic activities of the university without interfering in its affairs.

---

[34] She later became an eye specialist. Tan Chin Tuan's "Family Holding" and "Foundation" are well-known establishments in Singapore.

3

Professor of Chinese at Kuala Lumpur

Meanwhile the University of Malaya established a Department of Chinese Studies in Kuala Lumpur and invited Dr Cheng Te-k'un, Reader in Chinese Archaeology at Cambridge University to be its Visiting Professor to draw up a blueprint for the department.

In April 1963 my wife gave birth to our fourth daughter, Sook Pin. I was invited to give a public lecture in Kuala Lumpur at the University of Malaya. Before the lecture Rayson Huang, who was then the Dean of Science, briefed me on a list of names, suggesting that I should pay special attention to questions coming from them. I did not know the reason then, but later I realised that they were the names of certain members of the University Council on the selection board for the new Chair of Chinese Studies and that the lecture took the place of an interview for a position which I had not applied for. My personal particulars were sent to two external experts to be assessed. I never attempted to find out who they were through official sources, but later unsolicited information came to me from the two assessors themselves. One of them was Professor Patrick Fitzgerald of the Australian National University. I received an invitation from him to act as external examiner of one of his Ph.D. students as soon as I went to Kuala Lumpur. The other was Professor William Hung 洪煨蓮 of Harvard University. When I visited him in Cambridge, Massachusetts, in 1965, he advised me on book acquisition for the Chinese collection of the University of Malaya Library, telling me that there would be no point for the library to try to buy rare books. In September 1963 an invitation came from the University of Malaya, offering me the Chair of Chinese Studies.

I considered the pros and cons of moving to Kuala Lumpur. It was not an easy decision to make, but fortunately it was not a Hobson's choice — whatever my decision, the result would not be a disaster. Academically I saw a chair in Chinese giving me more flexibility in my research than a post in the history of science in the Physics Department. It would be very unlikely for a chair in history of science to be created in Singapore. It would also be unlikely for me to be appointed Professor of Physics in the university — and, even if appointed, the lesson I received in Cambridge on peerage ranking could not be erased from my mind and I would find it difficult to hold my own among my peers in physics. On the other hand, I would be the only professor of Chinese in the world specialising in the history of Chinese science, which would absolve me from competition with all the other scholars of Chinese. History soon proved that I had made the correct move. Three or four years after I left Singapore the new Vice-Chancellor, my friend Dr Toh Chin Chye, ordered the closure of the course on the history and philosophy of science at his university on the grounds of a small student number. Members of the Department of Philosophy were responsible for teaching the course at that time. I did not try to find out whether student number was the main reason, although I knew that my friend had problems with the Department of Philosophy in those days. Should he be hostile to the course itself, the history of science would not have much of a future in the University of Singapore.

One important factor to consider was acceptability, namely by the university, the government, colleagues in the department, the local community and the international community of scholars. An invitation from a university that had strong representation from the government already showed the acceptance by both the university and the government. I only needed to consider the last three groups.

I was getting on very well with my colleagues in the Physics Department in Singapore as physicists by training and as equals. In Kuala Lumpur I would have to deal with colleagues with different academic backgrounds. Besides being their administrative head, I would have to show academic leadership in research at the same time. I realised that if I were to carry on alone happily with my research on the history of Chinese science, I would be working in isolation, away from the research interest of my colleagues. I then had the idea of finding topics that involved science in literature, poetry, linguistics, etc., and to study them in collaboration with some of my colleagues.

Acceptability by the local community concerned a Chair in Chinese more than a post in physics or in the history of science. I had not received formal education in a Chinese medium school and I did not have any expertise in Chinese calligraphy, painting and poetry to impress the Chinese community, I was wondering how I would be received by the Chinese-educated in Malaysia. Fortunately I found three encouraging factors in my favour: (a) my good relations with the two major Chinese newspapers that included glowing reports about me as a scholar. My publications in these two newspapers showed that I could write in Chinese, (b) my contributions to the study of Chinese science, which had a special appeal to the Chinese-educated, together with my connection with Needham's work on the history of science in China, and (c) my friendship with Wang Gungwu's father, Mr Wang Fo-wen, who was then the most respected person in the Chinese-educated community in Malaysia. I thought out a strategy of getting in touch informally with the principals and teachers of the Chinese schools, either directly or through my colleagues, but without involving any sort of politics.

As for acceptance by the international community of scholars, I knew I would not stand a chance as a physicist, but perhaps as a historian of science, my chance would be better. As a Professor of Chinese Studies, I thought that I might be able to hold my own both as a scholar of Chinese and as a historian of science. In other words, Chinese studies would provide me a much wider playing field than the history of Chinese science itself. From the point of academic considerations the Chair of Chinese Studies seemed to me to hold better prospect. After considering everything, I then delivered my letter of resignation from my Readership in the University of Singapore to Dr B. R. Sreenivasan, who was also about to relinquish his own Vice-Chancellorship from the same university.

In April 1964 I assumed duty at the University of Malaya, Kuala Lumpur, as the first Professor and Head of the Department of Chinese Studies. I was fortunate to find the department being taken care of by the Acting Head, Professor Wolfgang Franke, from Hamburg University, who came to Kuala Lumpur as a Visiting Professor. Wolfgang Franke was one of the leading Sinologists in Europe and he was at the university to offer advice whenever needed. He came to visit the department quite regularly, and eventually, it turned out that he lived in Kuala Lumpur for a longer period than me. Cheng Te-k'un also visited the department during his sabbaticals in Cambridge. Hence the department was well-placed in maintaining Western traditions. I had to think of ways to satisfy the Chinese-educated community.

In 1965 I approached my Vice-Chancellor, Sir Alexander Oppenheim, to ask for funding to invite Professor Wang Shu-min 王淑岷 from the Academia Sinica to be the department's Visiting Professor for three years. Oppenheim was quite supportive, but he needed the support of the Dean of Arts, who was then the Professor of Indian Studies. He called for a meeting involving the three of us and asked the Dean for his opinion. With a smile the Dean said that it was a good idea, but added that the six other departments in the Arts Faculty, especially his own department, should each be allocated the post of a Visiting Professor as well. Oppenheim remained silent and looked towards my direction. I asked whether the Dean would raise the same issue if funding came from outside sources. He smiled again and answered that he would certainly not. The brief meeting ended and as soon as I got back to my department I wrote to Dr Dato' Lee Kong Chian about what happened and asked him for his help. He did not answer me directly, but the following week the Registrar received a letter from the Lee Foundation saying that it would be willing to provide funding for the appointment of Professor Wang Shu-min as Visiting Professor for as long as the University wished to retain his services. Oppenheim called another meeting of the three of us. After reading the letter from Lee Foundation with a smile, Oppenheim asked the Dean if he had any objection. "Certainly not" answered the Dean, also with a smile.

Wang Shu-min had been a Visiting Professor at the University of

Singapore. On one occasion he and I had lunch together with Dato' Lee Kong Chian at his Garden luncheon club. Later I also succeeded in inviting Dr Chien Mu 钱穆 to join the department as Visiting Professor. At one stage my department also obtained the help of the linguist Dr Jou Bienming 周辨明 to teach the Chinese romanisation system to beginners. At the suggestion of Wang Gungwu's father, Mr Wang Fo-wen, I appointed Chen Tieh-fan 陈铁凡 from Nanyang University as Senior Lecturer in the department. In 1965 I recruited Su Ying-hui 苏荧辉 from Taiwan to join our teaching staff. Before I assumed duty two members of the teaching staff were already on board. One of them was a Senior Lecturer in the person of Cheng Hsi 程曦, who later left to become a full professor at the University of Iowa. The other was a Lecturer, Chen Chi-yun 陈启云, who was then writing his Ph.D. dissertation for presentation to Harvard University and who later left to become a full professor at the University of California, Santa Barbara. Chen Hsi and Chen Chi-yun did not remain in the department very long. William Dolby, a research student of Van der Loon, and Dr Magdelene Dewall, a former research student of Cheng Te-k'un, came as replacements.

During my first few years in Kuala Lumpur, I believed that my department had a line-up of teaching staff that any university would be proud of. For example, Cheng Te-k'un designed a course entitled "Outline of Chinese Culture" that was taught at one time by an outstanding team — he was lecturing on Chinese archaeology, Wolfgang Franke on Bibliography and Ming history, Professor Harold Weins, Visiting Professor from Yale to the Geography Department, on Chinese geography, Chien Mu on neo-Confucianism, Wang Gungwu on overseas Chinese, Chen Chi-yun on Han China, Father Xaviour Thani Nayagam on Buddhism, and I on Daoism and Chinese science and technology. I was cautious about lectures concerning religion where the class included Muslim students, because of a law that prohibited the preaching of other religions to them. Thani Nayagam, the Professor of Indian Studies, was a Roman Catholic priest knowledgeable in the subject of Buddhism; no one could accuse him of preaching Buddhism when he lectured on the subject. I focused more on the philosophical side of Daoism, and mentioned that as a religion it accepted all others as being equally capable of leading to the Way. Professor Chen Shih-hsiang 陈世骧 of Berkeley, Professor Li Tien-yi 李田意 of Yale and Ohio and Professor Liu Ts'un-yan 柳存仁 had acted successively as departmental external examiners for the B.A. examinations during my term of office.

It was fortunate that the Ministry of Education under the Japanese government sent its Peace Corps Japanese language teachers to Kuala Lumpur to teach Japanese. As there was neither a Department of Japanese nor a language centre at the university at that time, I offered my department as the temporarily base for these teachers and to officially look after them. These teachers conducted Japanese language courses both inside and outside the university. From Wolfgang Franke I heard that students doing Chinese in leading American universities were required to gain a reading proficiency in Japanese. Accordingly I encouraged my students to pick Japanese as an optional subject. It was gratifying to see as their host that all the Japanese language teachers and my colleagues in the Department of Chinese Studies got on very well together.

As Professor and Head of Department, I felt that it was my responsibility to show academic leadership, so I needed to adapt the direction of my research activities to accommodate the new environment. For example, there was no longer any need to consider their proximity to physics, but instead their closeness to the Chinese culture should take priority. I tried to relate my research activities to the research interest of my colleagues. For example, I joined my colleagues Su Ying-hui and Chen Tieh-fan in the study of Daoist texts, with them looking after textual matters while I concentrated on matters pertaining to science. Likewise I studied the references to Chinese alchemy in the poems of Bai Juyi 白居易 and Lu You 陆游 together with Goh Thean Chye 吴天才 and the content of science in Li Ruzhen's 李汝珍 novel *Jinghuayuan* 镜化缘 with Yu Wang Luen 俞王纶. Another colleague, Ang Tian Se 洪天赐, asked me to suggest a topic for his Ph.D. thesis. I advised him to make a full study of Yixing's contributions to science in 8th-century Tang China. I wrote my publications mainly in English, but sometimes I wrote them in Chinese to avoid accusations that a Professor of Chinese of Chinese origin in a Chinese environment did not publish anything in Chinese, and also to avoid dissociating myself from colleagues who published solely in the Chinese language. I also began publishing in the Japanese language in Japanese journals.[35]

A Widening Horizon

When I was in Singapore I had visited England, Spain and the US solely on academic grounds, and had twice visited Hong Kong for social and recreational purposes. On my way to my destinations I had also visited France, Italy, Colombo and Bombay (now Mumbai) as a tourist. Like the proverbial frog leaving the well, those were the first few hops to see the outside world. My new appointment in Kuala Lampur had greatly widened my horizons. Soon after I assumed duty, the Dean of Arts, Robert Ho, nominated me as a university representative to attend a Leverhulme conference on extra-mural education held at the University of Hong Kong. That was in the month of October, when a number of important university functions were held. I was invited to several of them, including the ceremony to lay the foundation stone for the Robert Black College by the Governor of Hong Kong, Sir Robert Black himself. I could hardly dream that, 20 years later, I was to become the Master of that College.

Part of my duties involved dealing with foreign diplomats in Kuala Lumpur. On the American side, I assisted the Ford Foundation, the Asia Foundation, the Esso Foundation as well as the American-Malaysian Association to select candidates to receive financial support from them to study in the US. Likewise I offered my assistance to the Alliance Francaise. As mentioned earlier I provided a base for Japanese language teachers from the Peace Corps. Later the Japanese Ministry of Education sent out university teachers to replace them. The Japan Foundation also sent Visiting Professors to the Faculty of Arts, and I helped to make them feel welcomed. These visitors provided me a good opportunity to practise my Japanese, and I made many friends. Since Wolfgang Franke was a leading Sinologist in

---

[35] See Ho, P.Y. (1972), "Maraiya Daigaku ni okeru kagakushi no ici ni kansuru hōgoku", *Kagakushi kenkyū* マライ大学における科学史の位置に関する報告, 11:92–94.

Germany and many German Sinologists in those days had studied under his father, Professor Otto Franke. German Sinologists passing through Kuala Lumpur would make it a point to see him. He would always try to arrange for me to meet them as well. He was also a friend of the German ambassador in Kuala Lumpur. I thus had good contact with both German Sinologists and the German Embassy. In the 1960s the Republic of China had a Consul-General office in Kuala Lumpur, headed by Mr Chang Chung-jen 张仲仁. My colleagues from Taiwan maintained friendly relation with him. He sent his daughter to study for a degree in my department. Thus, on their national days, my wife and I were often invited to the respective embassy to join in the celebration.

Early in 1965 I received a Carnegie Corporation of New York travel grant to visit Yale University in New Haven, Connecticut. My former teacher, Professor Derek J. de Solla Price, who taught me applied mathematics at Raffles College, was then the Head of the Department of the History of Science and Medicine at Yale. Since I last met him in Ithaca in 1962, he had been trying to get me over to Yale for a period, pending the procurement of a travel grant. He arranged for a joint invitation from the Council of East Asian Studies under Professor Arthur Wright and the Department of History of Science and Medicine to me to visit Yale for a period of three months. Hearing about my visit to North America, Wolfgang Franke provided me with a list of Sinologists, whom I ought to call upon in order to make myself known. Accordingly, accompanied by my wife, we made Berkeley our first stop. There we called on Chen Shi-hsiang, who invited Professor Chao yuan-ren 赵元任 and Mrs Chao to join us at a Chinese restaurant for lunch. When I asked Chen Shi-hsiang whether he would be available to go to Kuala Lumpur as a Visiting Professor, he asked whether I would be free to visit Berkeley instead. We visited Stanford University and met Professor James Liu. Unfortunately he was too busy then and I did not have much of a chance to talk to him. The next time I saw him was more than 20 years later when we were guests of Professor John C.Y. Wong 黄靖宇 of Stanford. James Liu talked about his impending appointment with his medical doctor about his esophagus and left early. That was the second and last time I saw him. In Yale I met Arthur Wright, Professor Mary Wright, Professor Han

Frankel as well as Professor Li Tien-yi. I took the opportunity to ask Li Tien-yi to serve as the next external examiner when Chen Shi-hsiang's three-year term expired. During a visit to New York we went to Columbia University hoping to meet Professor Chiang Yee 蔣彝, of "Silent Traveller" fame after the name he used in his several books of travel and a friend of Rayson Huang. Chiang Yee happened to be away from his office, but Professor Hsia Chih-ching 夏志清 hosted us to lunch at the Golden Bamboo Restaurant on behalf of his colleague and friend.

In New York a friend, Mr P.W. Parsons, whom I first met in Cambridge, England, took us to the Harvard Club of New York for lunch. Professor Fred Hung (Hung Fu 洪绂) was then a Visiting Professor at the Geography Department of Yale University. He became our downstairs neighbour whom we often invited to join us for a meal. He took us in his new car to Cambridge, Massachusetts, to see his elder brother Professor William Hung (Hung Wei-lian). William Hung asked Professor Yang Lien-sheng 杨联升 to join us for dinner at his residence, but he was unable to attend as he had to play host to his visitor, Professor Li Chi 李济. I only managed to meet Yang Lien-sheng more than 20 years later in Hong Kong at the home of Cheng Te-k'un. We visited Nathan Sivin in Boston, and on a tour to Harvard University guided by a Malaysian friend Dr S.T. Leong (梁肇庭), we met Professor John Fairbank.

At Yale I gave a public lecture on the Chinese elixir of life. It was held at the Department of Anatomy. I introduced my lecture by asking why it was arranged to take place there. Should my host expect me to verify the power of the elixir, which the ancient Chinese alchemists believed to be able to resurrect the dead, by testing it on the cadaver kept nearby, I would have to offer my apologies for not having any elixir with me. At the Department of he History of Science and Medicine I joined Dr Bernard Goldstein in writing a paper on the 1006 supernova for *The Astronomical Journal*.[36]

The new Vice-Chancellor of the University of Singapore, Lim Tay Boh, asked me to find out from Dr Nelson Wu of Yale University whether he would

[36] See Goldstein, B.R. and Ho, P.Y. (1965), "The 1006 Supernova in Far Eastern Sources", *The Astronomical Journal*, 70.9:748–753.

consider accepting an offer from Singapore to be the Curator of the Museum of Art in his university. I got Li Tien-yi to introduce me to Nelson Wu. As that was the first time we met, I did not raise the question of Singapore but arranged to see each other again in the near future. Then news came that two senior members of the staff in Yale had lost their tenure, and one of them was Nelson Wu. Thinking that my chance had come, I went to see Li Tien-yi and told him my intention to speak to Nelson Wu. Li Tien-yi, being a close friend of his, told me that he was considering the 11 offers that he had already received. Knowing that my friend Lim Tay Boh would stand no chance in competing against American universities and museums, I aborted my attempt to see Nelson Wu. I was already aware of the geographical vastness of North America; my American schoolteacher had taught me about thinking big, because there were other countries in the world with vast areas. At Berkeley I was amazed to find out that the budget for the Lawrence Laboratory alone far exceeded that of the State of Singapore, and here there were the large number of academic openings available for one to choose. I had broadened my horizons of the outside world.

On the personal side my third cousin, Paul S. Ho (何兆中), a Professor in Electronics at Cornell University, visited us and stayed overnight at our apartment at Yale. Paul later joined IBM and then moved to Houston. In New York City we met Bevan E. Foo (Foo Eng Lan), a former student of mine in St. Michael's Institution in Ipoh, who was then working with Pan American Airlines. He and his wife took us to Chinatown for lunch. When we left New York on a TWA flight to San Francisco we were surprised to find Bevan coming on board. When I asked him why he took a TWA and not a PanAM flight, he answered that he came only to bid us farewell and that as a ground engineer he was allowed to board the plane.

We returned to Kuala Lumpur by way of Japan. Although I stayed overnight at the Nikko Hotel in Ginza *en route* to Hawaii and Los Angeles, that was only a scheduled stopover of Japan Air Lines and I only got to see Tokyo through the bus window late at night and early in the morning. My first real visit to Japan was in 1965. In Tokyo I searched for ancient Japanese astronomical records with the help of Shigeru Nakayama. In Kyoto I visited the Jimbun kagaku kenkyusho of Kyoto University and was invited to dinner by Kiyosi Yabuuti at the Kyoto Hotel. There I met all the senior members of that Institute. We then went to Kobe and Osaka. In Kobe we visited my old Japanese teacher, Professor Hiroshi Suguro at Kobe University. He arranged for his former colleagues in Ipoh to meet us. It was the first time that the group met since they were repatriated home after the war. My visit initiated their annual gathering under the name *Ippōkai* 一步会, literally meaning "One-Step Meeting", but also embodying the original purpose of "Meeting of former Ipoh colleagues". Annual meetings, where members dined together in a holiday resort and talked about old times, continued until the late 1990's when old age relentlessly took its toll on the members. I had attended as a guest at least half a dozen of their meetings.

The University of Malaya being new, had a shortage of staff quarters. The university rented a house for us, but it was inadequate for my family of five children and my books. It fell far short of House number 8, University, where we used to live in Singapore. Instead of waiting for the availability of proper university quarters, I decided that it was time to own a house we could call home and bought a house-to-be-built off the design from the Petaling Garden Estate. The architect for the housing estate, Lee Hong Kai, told me that I was once his teacher at St. Michael's Institution, Ipoh. Trusting that the construction of our first house was in good hands, my wife and I set off for North America, and upon our return to Kuala Lumpur it was there ready for occupation.

Shortly after moving into our new house I had to make a trip, again accompanied by my wife, to Hong Kong and Taiwan to interview candidates who had applied for a post of lectureship in the Department of Chinese Studies at my university. Our plane landed in Hong Kong at a time of civil commotion, and we were escorted to our hotel under guard. Professor Lo Hsiang-lin 罗香林, who had recently been appointed Head of the Department of Chinese at the University of Hong Kong, let me use one of his departmental offices for interviewing the candidates.

We received an official reception from the Ministry of Education upon our arrival at Taipei. The Ministry also arranged for me to interview candidates at the Taipei Medical School. There was also an invitation to dinner from the Minister Yen Chen-hsing 阎振兴. Doubtlessly the Consul-General in Kuala Lumpur, Chang Chung-ren, had made all these arrangements for us. Su Ying-hui, a classical scholar and specialist in Dunhuang studies, was the successful candidate for the post in Kuala Lumpur. We visited the Academia Sinica in Nankang, where Wang Shu-min introduced us to the President, Dr Chien Ssu-liang 钱思亮. We also met the Korean scholar Professor Cha Chu Whan 车柱环, a specialist in the poet Tao Yuanming 陶渊明. Wang Shu-min invited us to dinner at the Ermei Restaurant which served Siquan food. He also invited his colleagues at the Chinese Department of the National Taiwan University to join the party. There we met some of the leading traditional Chinese scholars, including Professor Mao Tze-shui 毛子水, Professor Tai Ching-nung 台静农, Professor Chang Ching 张敬, Professor Chu Wan-li 屈万里 and Professor Li Hsiao-ting 李孝定. On the personal side, I visited the headquarters of the Radio Broadcasting Station to pay my respects to my father's old friend, Mr Liang Han-chao 梁寒操, who had acted as best man during my parents' marriage in China. My wife and I also took a brief holiday at the Sun Moon Lake, visiting Taichung along the way.

I first found myself being dragged into university politics in 1965. A seat in the University Council for a Senate representative fell vacant. A member in Senate, who once stood as a candidate for the post of Vice-Chancellor, was in the running, but some of the professors tried to stop him by putting me up as a candidate, thinking that my connections with the three Faculties of Arts, Science and Medicine would come in useful. I won with two-thirds of the votes and became a Member of Council for a term of three years.

When I reported to Lim Tay Boh, the Vice-Chancellor of the University of Singapore, on the outcome of my meeting with Nelson Wu in Yale, he had another proposition to make. The Head of Department of the Chinese Department at his university had retired, and he wished me to return to Singapore to become the Professor and Head of that Department. Being personal friends, I could tell him frankly that if I were to help him to solve his problem in this manner, I would create a new problem for my university in Kuala Lumpur. I suggested to him the names of three scholars whose knowledge of Chinese was better than mine and also advised him on the choice of two assessors, one was Chien Mu to satisfy the Chinese-educated and the other was Denis Twitchett to satisfy the Sinologists. He eventually decided to invite Jao Tsung-i to take up the Chair of Chinese in the University of Singapore.

Oppenheim retired as Vice-Chancellor of the University of Malaya in 1966 and Rayson Huang was appointed Acting Vice-Chancellor for a one-year period. Rayson Huang appointed me to act for him as Chairman of the Joint Publication Committee for the University of Malaya Press and the Malaysian Branch of the Oxford University Press. The two presses produced a number of academic publications mainly concerning Southeast Asia. Rayson Huang also made me deputise for him as member of the Loke Yew Scholarship Board, but the Board had never met. Before the formation of ASEAN, the three countries of Malaysia, Thailand and the Philippines got together to form an Association of Southeast Asia — ASA for short. The Malaysian government appointed me as one of its representatives for the second ASA education conference held in Bangkok.

The term of Professor Xaviour Thani Nayagam as the Dean of Arts was to expire in 1967. At the early stage of the University of Malaya few professors would eye this post. Wang Gungwu was the first Dean of Arts in the university in 1962, but he relinquished the post when he was appointed to the Chair of History. The Deanship went to Robert Ho, Professor of Geography, who used to refer to himself as the top office boy in the Faculty. Robert Ho did not remain Dean for long; he left Kuala Lumpur to become Professor of Geography at the National Australian University, Canberra. His successor was Thani Nayagam. One evening three professorial colleagues from the Faculty of Arts took me out for dinner at a Chinese restaurant in Kuala Lumpur where they persuaded me to take up the Deanship. By then Needham had already been asking me when I would return to Cambridge. I promised to take it up for only one year, saying that my sabbatical leave would be due in 1968 for my return to Cambridge.

In April 1967 I became the Dean of Arts. Since I became an ex-officio Member of the University Council, I had relinquished my post as Senate representative. A new Vice-Chancellor had arrived in the person of Dr James Griffiths from Magdalen College, Oxford. His term of appointment was only for one year, as the University hoped to find a local candidate to fill that office. One of his early meetings with the deans concerned a state visit to Malaysia by President Ferdinand Marcos of the Philippines. In a letter from the Prime Minister, Tunku Abdul Rahman in his capacity as Chancellor of the University, the University was asked to consider conferring an Honorary Degree on the visitor to reciprocate the Honorary Degree that the Prime Minister received during his state visit to the Philippines. Dr Griffiths said that although the letter came as a request one had to take it as an order, whereby he requested the Dean of Arts, in the same vein as the letter, to be the Public Orator for the conferment of an Honorary Degree of Doctor of Law on President Marcos. I performed my duty accordingly. I went to Manila, representing the Malaysian government in the third ASA education conference. That was the last conference of that nature; ASA was soon to be replaced by ASEAN.

The Asian and North African Congress would hold its international conference in Ann Arbor at the University of Michigan in the summer of 1967. While I was sitting in the Dean's office contemplating applying to the university for a travel grant to enable me to attend the conference, a telephone call came from the Vice-Chancellor's office about a message from the American Embassy. The message said that the embassy had some excess funds for the year, and if the Dean of Arts had any intention to visit the US he would be welcomed as a guest. It looked as if someone could read my mind over there. I accepted the offer spontaneously and visited North America under the Leaders and Specialists Programme. My first stop was Washington, DC, where the State Department arranged for me to stay at the Windsor Hotel. I called at the State Department for a briefing and to draw up a plan of my visit. I said that the main purpose of my visit was to attend the conference in Michigan, and that a friend had reserved a room for me at the Harvard Club of New York to enable me to stop over and discuss an academic paper on Chinese astronomy. While I was talking to my host, he received a message that there was a young lady waiting outside to see me. My host asked me who she was. I said that she was an American scholar, who had come to see me in Kuala Lumpur while doing research on Chinese women in Malaysia. After introducing my visitor to my host, I said, "Her interest is in Chinese women." With a smile she added, "Only academically." My host then asked her what she was doing. She answered that she just got back from her research in Malaysia and would soon be looking for a job. My host then asked her to see him afterwards for an interview for a post in his department. It was a pleasant surprise to find things could happen this way.

The State Department arranged for me to visit the Smithsonian Institute before I left Washington, DC. In New York City my friend brought me to the New York City Library to write a paper on the 1054 Crab Nebula together with a librarian of that Library. The paper later appeared in *Vistas in Astronomy*.[37] As it was the State Department that arranged for me to attend the conference in Ann Arbor, my name tag carried Washington, DC as my address. I delivered a paper on astronomy in Ming China. The editor of the *Journal of Asian History* was present during my presentation. He asked for a copy of my delivery for publication in his journal.[38]

The State Department also provided me a return air ticket to Montreal to view the Exhibition. However, shortage of time prevented me from visiting Canada and I had to return the unused tickets to Washington. I met a large number of eminent scholars of Asian studies at the conference and also some old friends. I wrote a report to the American ambassador on my visit upon my return to Kuala Lumpur. In the same year of 1966 I contributed five articles to the *Dictionary of Scientific Biography*, edited by Professor Charles Gillespie of Princeton.[39] One of these articles was selected for special mention in a review of the *DSB* in the *New Yorker*.[40]

---

[37] See Ho, P.Y., Paar, F.W. and Parsons, P.W. (1971), "The Chinese Guest Star of 1054 and the Crab Nebula", *Vistas in Astronomy*, 13:1–13.

[38] See Ho, P.Y. (1969), "The Astronomical Bureau in Ming China", *Journal of Asian History*, 3.2:135–157.

[39] See Ho, P.Y. (1971), "Ch'in Chiu-shao, 13th-Century Chinese Mathematician", *Dictionary of Scientific Biography*, American Council of Learned Societies (Baltimore), 3:249–256; Ho, P.Y. (1971), "Chu Shih-chieh, 13th-Century Chinese Mathematician", *Dictionary of Scientific Biography*, *ibid.* 3:265–271; Ho, P.Y. (1973), "Li Chih, 13th-Century Chinese Mathematician", *ibid.* 8:313–320; Ho, P.Y. (1973), "Liu Hui, 13th-Century Chinese Mathematician", *ibid.* 8: 418–424, and Ho, P.Y. (1976), "Yang Hui, 13th-Century Chinese Mathematician", *ibid.* 14:538–546.

[40] June 16, 1980, 117–121, esp. pp.119–120.

In January 1968 I attended a symposium on the History of Chinese Science organised by Dr George Wong at the Chinese University of Hong Kong and presented a paper on chemical information contained in the Chinese pharmacopoeias.[41] In the 1960s the number of scholars engaged in research into the history of Chinese science was not that many. Needham

```
        SECRET


                    UNIVERSITY   OF TY IALAYA

                                                    Pantai Valley,
                                                    Kuala Lumpur.


                                                    March 25, 1968.



        Professor Ho Peng Yoke,
        Department of Chinese Studies,
        University of Malaya.


        Dear Professor Ho,

                The Committee appointed by the Chancellor of the
        University of Malaya to recommend a suitable person for the
        appointment of Vice-Chancellor, has been considering several
        nominations and I am pleased to inform you that your name is among
        those being considered.   While the Committee appreciates that
        some of those nominated may not wish to be considered for appoint-
        ment, we in the Committee nevertheless felt it necessary to have the
        curriculum vitae of all those who have been nominated.

                I shall be grateful therefore if you will assist the
        Committee by providing your curriculum vitae on the lines of the
        attached form.   It should be sent to the Registrar, University of
        Malaya, Pantai Valley, Kuala Lumpur.

                Your assistance will be much appreciated by my Committee.


                                            Yours sincerely,

                                            Ismail Md Ali

                                                Chairman,
                                       Vice-Chancellorship Committee,
                                         University of Malaya.
```

Figure 7 Letter from Chairman, Vice-Chancellorship Committee

---

[41] See Ho, P.Y. (1958), "Alchemy of Stones and Minerals in Chinese Pharmacopoeias", *The Chung Chi Journal*, 7:155–170.

and the few historians of Chinese science in China were not present at this conference. Kiyosi Yabuuti, Shigeru Nakayama and Nathan Sivin were there.

The term of office of Dr Griffiths as Vice-Chancellor of the University of Malaya expired in 1968. The University Council set up a Search Committee, with Tan Sri Ismail Ali as Chairman, to look for a new Vice-

---

UNIVERSITY OF WASHINGTON
SEATTLE, WASHINGTON 98105
26 December 1967

Far Eastern and Russian Institute

Professor Ho Peng Yoke
University of Malaya
Pantai Valley
Kuala Lumpur, Malaya

Dear Professor Ho:

    I wanted to discuss with you the feasibility of inviting you to join our Institute for a year on a guest professorship arrangement. We have such a position available for the academic year 1968-69, and if this idea fits into your general inclinations and your time schedule we would feel greatly honored if you would be willing to share your expertise and your scholarship with us and our students.

    Under this arrangement you would be expected to give during all three quarters (our University is running on the quarter system) a reading seminar for advanced graduate students. This seminar is not so much meant to elucidate linguistic and philological problems; the students are expected to have already a certain mastery of these questions when they come to you. It is rather tailored toward exposing the students to a wider variety of texts and the focus would be on the content of these texts. The choice of the texts would be yours. Generally they have been taken in the past (Professor Hsiao Kung-ch'uan was the one who used to give this course) from the field of intellectual history.

    In addition we would very much like you to give each quarter a course or a seminar pertaining to your special field of interest. For these we would expect your proposals so that they can be put on the program. They might include, if you so choose, one course for advanced undergraduates not necessarily only for those in Chinese studies (that is to say without the language requirement) on the History of Chinese Science.

    If you are interested I would be happy to have your curriculum vitae. I look forward to hearing from you.

Sincerely yours,

George E. Taylor
Director

Figure 8 Letter from Seattle

Chancellor. I received a letter of invitation from the Chairman asking me to submit my curriculum vitae for consideration by the Board. I thanked him for noticing me, but requested that he did not need to proceed with my case any further. I was about to apply for my sabbatical to return to Cambridge. Following the retirement of Professor K.C. Hsiao 蕭公权 at the University of Washington, I received an invitation from Seattle to teach the history of ideas, with an emphasis on scientific ideas, as a temporary measure to fill the vacancy. I sent my apologies and declined the offer because of my earlier commitment to Needham.

Second Visit to Cambridge

Meanwhile Needham kept asking me to return to Cambridge, and in one of his letters he said that he was waiting for the definite date of my arrival so that he could plan to put things aside in order to work together with me on the final drafts of the alchemy section of *Science and Civilisation in China*, volume 5. By April 1968 I had already put in four years of service in my university and was entitled to take eight months' sabbatical on full salary with

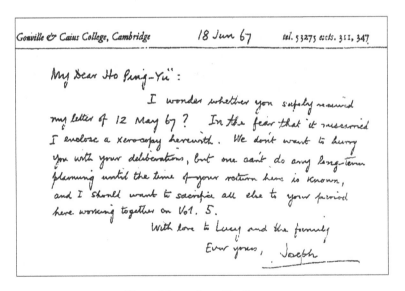

Figure 9 Letter from Needham

an overseas living allowance and return air travel expenses. I relinquished my Deanship of the Faculty of Arts and proceeded to Cambridge. Needham was then the Master of Gonville and Caius College, and lived in the Master Lodge. He put me up in a room at the west wing of the Lodge. We worked together for about a month and completed the final drafts of what finally turned out to be the two sub-volumes Part 3 and Part 4 of volume 5.

By then Needham had exceeded the retirement age for Cambridge University, although a Master of a College need not abide by this regulation and could serve a few years longer. He thought ahead and told me that if Dorothy Needham were to survive him his books would impose a heavy tax burden on her. Accordingly he formed a Trust to take possession of his books. I witnessed the signature of one of the five trustees, Dr George Salt, FRS. The other four were Peter George Burbidge, Lu Gwei-Djen, Eric Betrand Ceadel and Walter Brian Harland. This was the modest beginning of the Needham Research Institute Trust. Burbidge, the Production Manager of Cambridge University Press, which was responsible for bringing out the *Science and Civilisation in China* series, and a trusted friend of Needham, was the Chairman of the Trust.

I was away from Cambridge for three days visiting West Germany. Wolfgang Franke on behalf of the Chinese Department and the Astronomy Department of Hamburg University invited me to Hamburg to give a lecture. I also contributed two articles to the *China Handbuch*.[42]

My wife visited me in Cambridge together with our son Yik Hong and our third daughter Sook Kee during their school holidays in July and August. They stayed together with me at the Lodge and we often had our meals together with the Needhams and Lu Gwei-Djen. One evening Needham poured each of us a glass of wine and said, "Let us rise to toast to the new *yangguizi* 洋鬼子 (foreign devil) of Britain." He then added that Lu Gwei-Djen had become a British citizen and that she often teased him with that title. On another occasion Lu Gwei-Djen remarked that there were three Masters

---

[42] See Ho, P. Y. (1974), "Kalendar und Datierung", *China-Handbuch* (Hamburg), pp. 602–606 and Ho, P.Y. (1974), "Astronomie", *ibid.* pp. 63–64.

around — one was Needham, another was my son and the third was the dish master in the kitchen.

Needham no longer had his Armstrong-Siddley convertible, having changed it for a blue Ford Contina. Lu Gwei-Djen told me that he did not like the original blue colour of the car and had it repainted to match the Chinese porcelain blue, known as *jingtailan* 景泰蓝, named after the source of production that made Chinese porcelain so famous. The final result was not a perfect match, but Needham had to make do with it. I was partly sad and partly amused when I saw Needham locking up his new car. The last time I was in Cambridge, when Needham saw me locking up my car, he told me that Cambridge people did not do such a thing. I replied saying that I greatly admired and respected the honesty of the people in Cambridge, but if I were to get into their habit of leaving my car unlocked, I would lose it very quickly when I returned to my own country. When I saw Needham locking up his Ford Cortina, I felt sorry that Cambridge had changed. Ten years ago when we left for a tour in Scotland, we had absent-mindedly left our door key in the keyhole. We returned after 10 days — everything was intact, including my wife's diamond wedding ring which was left on the table in the sitting room. Newspaper vendors then would leave their papers in the verandah, leaving some money behind for customers to make payment and collect their change. We were deeply impressed by the peaceful conditions of the town and the honesty of the people.

There was a Daoism conference to be held at Villa Serbollini in Lake Bellagio, Italy. The organiser of this conference was originally Helmut Wilhelm from the University of Washington, but because of ill-health Arthur Wright took over. Both Nathan Sivin and I had received invitations to attend. Needham had not received any, but he seemed keenly interested in the conference. Arthur Wright and Charles Gillespie were the two major critics of Needham's work, but Needham had not met either of them before. Needham hesitated to write to Arthur Wright, unsure if he would be welcomed. I wrote quietly to Nathan Sivin, asking him to speak to Arthur Wright. An invitation to Needham from Arthur Wright soon followed. When the two of them met later in Italy, they greeted each other like old friends meeting after a long absence. Nathan Sivin and I looked on with a smile that would be better expressed by the Chinese phrase *huixin weixiao* 会心微

笑 — a gentle smile arising from mutual understanding of the issue. Needham was unaware of what we did and did not take any notice of us.

Although Arthur Wright was one of Needham's major critics, at this conference they became great friends. Upon his return to Yale Arthur Wright decided to raise funds in the US to support Needham's work. Dr Thomas H. C. Lee 李弘祺 informed me that when he visited his teacher Arthur Wright at Yale in the summer of 1976, he heard him talking about raising funds to support Needham. Unfortunately a few days later Arthur Wright collapsed while playing golf and died. Dr Cliff A. Shillinglaw of Coca-Cola Inc. established the US Branch of the East Asian History of Science Foundation in 1977. However, it was no longer possible to find another Needham supporter like Arthur Wright, who was a much-respected scholar with substantial financial means to enable him to rub shoulders with the wealthy in the United States. For example, he only needed to speak to the Rockefeller Foundation to obtain the use of Villa Serbollini Bellagio for the conference on Daoism. His colleague as well as a mutual friend of Needham, Derek de Solla Price, joined the US Branch for a few years until his untimely death in 1983.

Before the Daoism conference, I had to attend the 12th International Congress held in Paris first. My wife and our two children accompanied me to Paris, where we stayed in the Hôtel de France et Choiseul along rue Saint-Hororé, the same hotel as the Needhams and Lu Gwei-Djen. While Needham and I attended the Paris meeting, Dorothy Needam and Lu Gwei-Djen enjoyed themselves playing Scrabble, sometimes watched by my children. Dorothy Needham and Gwei-Djen often played Scrabble together. In Cambridge Dorothy's sister, Muriel Moyle, sometimes joined them, but not Needham. Lu Gwei-Djen said that Needham did try, but because of his inclination to form long words he lost the game to his opponents who gained more points from short words that were strategically placed.

I found myself in a rather awkward position when I went to the session at which Needham was to present his paper. The chairman for the session was absent, and Needham proposed that I take the chair, which was unanimously agreed. Needham presented his paper, but intruded into the time slot for the next speaker. I indicated to him that his time was up, but it was no avail. I sat there anxiously, not knowing what to do. Needham ended his delivery only after he had used up more than twice his allocated time. He received warm applause from his audience. This incident which caused me great anxiety turned out to be a blessing; the next two speakers did not show up to read their papers. When my turn came to present my paper, I spoke on alchemy in Ming China.[43] I chose the Ming period because of the interest and expertise of my colleague Wolfgang Franke in that particular period of Chinese history.

After the conference in Paris I took my family by car to Cannes and then through the southern coast of France to Italy until we reached Rome. From Rome I saw my family off at the airport, as they had to be back in Kuala Lumpur for the new school term. I then took a train to Milan, where a car from the Villa Serbonilli picked me up for the venue of the Daoism conference. The villa was a property of the Rockefeller Foundation, which acted as our host. Needham attended this conference with his wife, while Lu Gwei-Djen went to London to visit the family of Professor Benjamin Platt, who was the head of the Henry Lester Medical Institute in Shanghai, where she worked before she went to Cambridge in 1937 to do a doctoral degree under the supervision of Dorothy Needham.[44]

1969 Edward Hume Lecture at Yale

Having completed the drafts for the alchemy sub-section, I returned to Kuala Lumpur with the feeling that my collaboration with Needham had come to a happy ending. I did not know then that Needham had other things in mind for me. Back in Kuala Lumpur I received a second letter from the University of Washington asking me to reconsider its offer. Having already used up all my sabbatical leave, there was no way for me to leave the univer-

---

[43] See Ho, P.Y. (1971), "Alchemy in Ming China", *Actes XIIth International Congress of History of Science* (Paris), 3A:119–123.

[44] Benjamin Platt was back in London as a professor in the School of Tropical Medicine. One of his students, Dr Cartwright, became the family doctor of Needham and Lu Gwei-Djen during the last years of their lives.

sity for any long period of time. Reluctantly I had to decline once again the offer from Seattle.

The University of Malaya had already appointed a new Vice-Chancellor in the person of Professor Ungku Aziz, formerly the Dean of the Faculty of Economics. A pleasant duty that he enjoyed in his early career as Vice-Chancellor was perhaps to play host to Crown Prince Akihito and Princess Michiko of Japan when the royal couple visited the University of Malaya in the afternoon of 20 February 1969. As a holder of a doctoral degree from Waseda University, Ungku Aziz was able to converse fluently in Japanese with his guests. I sat nearby behind them during the reception ceremony, where students from the university sang a number of Japanese songs. When they sang the song *yashi no mi* 椰の实 (the cocoanut) Princess Michiko discarded protocol and walked towards the students to join them in their singing. That surprised and delighted everyone.

I was re-elected member of the University Council as a representative of the Senate. Wang Gungwu had left the university to succeed C.P. Fitzgerald as Professor and Head of the Department of Far Eastern History in the School of Pacific Studies at the Australian National University. Rayson Huang, who was then the Professor of Chemistry, was the Chairman of the Technical Sub-Committee of the new Tunku Abdul Rahman College which the founders and organisers hoped would become a university in the future. It took 35 years for their dream to be realised when the Tunku Rahman University was established in 2003. Rayson Huang got me into his committee as a member, but he soon resigned to become the Vice-Chancellor of Nanyang University in Singapore. At his suggestion I succeeded him as Chairman of the Technical Sub-Committee. Shortly afterwards the College offered me the post of Principal. Negotiations between the College and the University of Malaya resulted in a suggestion by the Vice-Chancellor that I could retain my post as Professor and Head of Department in the university, with the freedom to assist the College under the title of an Advisor, but not Principal. The College compromised by giving me the title of Director and provided me with a car and chauffeur together with a monthly allowance to compensate for the time loss resulting from my involvement with the College.

Yale University invited me to be the 1969 Edward Hume Lecturer. I left in late April for America by way of London. It so happened that there were two applicants from England for a vacancy in my Department of Chinese Studies. It would save my time and university money if I interviewed them on my way taking advantage of the airfare provided by Yale. I also interviewed an applicant in London for the Chinese Librarian post on behalf of my university Librarian. Lu Gwei-Djen underwent a surgical operation and I would like to make a trip to Cambridge to visit her as well. Knowing this Rayson Huang asked me to drop by Peterhouse College to see someone who had been offered an appointment in his Chemistry Department, to find out when he could go to Kuala Lumpur. I accomplished all these missions, except that the candidate that Kuala Lumpur hoped to get invited me to lunch at Peterhouse and told me that Lee Kuan Yew, the Prime Minister of Singapore, had come personally to Cambridge to ask him to take a job back in Singapore. When I reported this back to Rayson Huang later, he accepted the facts of life with a laugh saying that he could not compete against Lee Kuan Yew.

I arrived in New Haven, Connecticut, on the 1st of May. Mary Wright took me to see a film on China by Edgar Snow. The next day I delivered my lecture on the *Book of Changes* and Chinese science. I did not know then the high prestige the Edward Hume Lecture enjoyed in the circles of Asian Studies in North America until some 20 years later when Ho Ping-ti, who gave the lecture in 1970, told me so. At Yale I met my former teacher, Derek de Solla Price. He told me that he and his family would soon visit Kuala Lumpur. I promised to find them accommodation in the University of Malaya and to arrange a welcome dinner reception at a Chinese restaurant. I could only be in America for three days, because of an impending general election in Malaysia on the second Saturday of May. The next day my host, Arthur Wright, asked his colleague K.C. Chang (Chang Kuang-chih 张光直) to take me in his car to the bus terminal to catch a limousine for the Kennedy International Airport.

I stopped over briefly in Tokyo, where Shigeru Nakayama took me to see Eri Yagi 江里八木, a former research student of Derek de Solla Price at Yale, and now a Professor at the Tōyō University of Engineering. She was then at a symposium on the history of science. She asked me to give a talk at the meeting, telling the audience about the lecture I gave at Yale. She said that as the organiser of the symposium she could make changes to the programme. She gave me the time slot originally meant for herself by telling the audience that it was a rare occasion for them to listen to me, whereas they had ample opportunities to hear her speak. I spoke in Japanese and gave the original text of my lecture to Eri Yagi for publication in a Japanese journal.[45] That evening I called at the Imperial Hotel to pay a social call to the newly arrived Ambassador of Singapore to Japan, Professor Ang Kok Peng 洪国平. Kok Peng and I were college mates in Singapore and the two of us shared the same university quarters during the early part of our teaching careers. He was staying at the hotel temporarily while waiting to move into his official residence.

Before I left Kuala Lumpur for England and the US the Ambassador of South Korea asked me to help him find out the best way for his son to pursue higher education in Britain or the US. He had left Kuala Lumpur to return to South Korea as the new Foreign Minister when I left on my trip to Yale. I consulted some friends in British and American universities on the issue and made an overnight stopover at Seoul to report to him.[46] Figure 10 shows a letter written later by Choi Kyu Hah 崔圭夏 after he became Prime Minister of South Korea. The letter talks about his son.

Communal Disruption in Malaysia on 13 May

On my return to Kuala Lumpur I started making arrangements to welcome Derek J. de Solla Price and his family. I found him a fellow's flat in the Second Residential College of the University of Malaya, where Professor Thong Saw Pak, the Master of that College, was also a former student of his. I also reserved a table at a Peking-style restaurant for the evening of 13 May

---

[45] See Ho, P.Y. (1972), "The System of the *Book of Changes* and Chinese Science", *Japanese Studies in the History of Science*, 11:23–39.

[46] His son took Chinese Studies as a subject at the University of Malaya. From Foreign Minister he later became the Prime Minister of South Korea. He was also briefly the President of his country.

REPUBLIC OF KOREA
OFFICE OF THE PRIME MINISTER

January 12, 1976

Dear Professor Ho:

    Thank you for your congratulations on my
appointment as Prime Minister of the Republic of
Korea.

    A bit of news about my son, Yoon Hong, who
was once your student at the University of Malaya.
With completion of his three-year military service,
he has now turned out to be a grown-up, and married,
and been employed at the Korea Trade Promotion Corporation here in Seoul.

    Wishing you good health and continued success
in your academic endeavors, I remain

                        Sincerely yours,

                        CHOI, KYU HAH
                        Prime Minister

Professor Ho Peng-yoke
Chairman, School of Modern Asian Studies
Griffith University
Nathan, Queensland 4111, Australia

Figure 10 Letter from South Korean Prime Minister

and invited Thong Saw Pak and his wife as guests so that they could come together with the Prices. The Prices arrived on 12 May and were fetched from the airport by Thong Saw Pak. I was planning to meet them the following day. The next day came and after I returned from work I went with Lucy to do some grocery shopping before getting ourselves ready to go to the restaurant. While shopping we heard over the radio that racial riots had erupted and the government had declared a state of emergency for the whole country. We could not go to the restaurant; Thong Saw Pak arranged for a special escort to take the Prices to the airport at Subang.

The general election saw the ruling government returning to power, but with its majority in the Lower House reduced. At that time the coalition government, the Alliance, consisted of the United Malay Nationals Organisation (UMNO) Party, the Malaysian Chinese Association (MCA) Party and the Malaysian Indian Congress (MIC) Party, with UMNO as the dominant party. The two minor parties suffered more casualties than UMNO. Politics sometimes have a direct bearing on the academic atmosphere in a university. Towards the later years of the 1960s Malays and the original natives of the country, together known as *bumiputra*, made up more than half the population of Malaysia, Malaysian Chinese made up about one-third, while the rest of the population consisted of Indians, Eurasians and others. While political power was in the hands of the Malays, the Chinese controlled much of the wealth in the private sector. Hundreds of people lost their lives to the riots; most of the casualties were Chinese. The Deputy Prime Minister, Tun Abdul Razak, declared a state of emergency and suspended the Senate and the House of Representatives. The National Operations Council, with Tun Razak as the Director of Operations, took over the day-to-day running of the country. The National Consultative Council took the place of the two Houses, but all members were appointed with no voting rights. They were free to discuss anything, except to propose a motion for voting. There was not supposed to be any showing of hands. The National Consultative Council started off with 65 members, government and unofficial. Thirteen of the unofficial members were representatives of professional bodies and two of them represented institutions of higher learning in the whole country.

The Vice-Chancellor of the University of Malaya, Ungku Aziz, and I were the two representatives of institutions of higher learning. The main theme of discussion was how to bring about national unity and to find ways and means to improve the economic standard of the *bumiputra*. I spoke on several occasions, pointing out that the races in the country were all inter-dependent, and that while planning to improve the lot of the *bumiputra*, other races in the country should not be totally neglected. I gave the example of sharing a cake. Instead of giving the *bumiputra* a larger slice out of the same cake at the expense of the other races, one should plan to make a larger cake — let the *bumiputra* have a larger piece and the others retain at least their former share, if not even something a little larger. It was important to cultivate goodwill among the different races. A Goodwill Committee, headed by Dato' Hussein Onn, was formed within the National Consultative Council, and I was appointed a member of this committee. As mutual understanding of one another's customs would help to promote goodwill, I wrote about the religions, taboos, popular beliefs and festivals of the Chinese. We encouraged different races to join their friends in their festivals.

In the same year I received the higher doctoral degree of Doctor of Science from the University of Singapore. The situation that Rayson Huang described as a rarity concerning higher doctoral degrees among university staff in Singapore in the early 1950s had slightly improved by then. Nevertheless, Ungku Aziz and Rayson Huang both wrote to congratulate me; the former regarded it as an honour to his university. However, my higher doctoral degree in science came shortly before my academic link with physics finally came to an end. When I took up the appointment as Professor of Chinese Studies at the University of Malaya in April 1964 my professional career as a physicist ceased automatically, but my connection with physics lingered on.

Towards the later part of the 1960s, astronomers and astrophysicists were much excited by the discovery of pulsars, and some thought that ancient astronomical records might help to discover new ones. My publication on Chinese records of comets and novae turned up handy. The Institute of Physics in London asked me to contribute an article on the applications of ancient Chinese astronomical observations for publication in its *Physics Bulletin*.[47] I was later elected Fellow of the Institute of Physics. A colleague of mine in the Faculty of Arts, Tan Sri Hamzah Sendut, became the Vice-Chancellor of the new Universiti Sains Malaysia in Penang. He asked me to join him as Professor of Physics in his university. Remembering my Cambridge experience I told him frankly that although I might be able to run the department and perhaps teach a few courses in physics, I would not be

---

[47] See Ho, P.Y. (1970), "Ancient Chinese Astronomical Records and their Modern Applications", *Physics Bulletin,* 21:260–263.

able to bring credit to his university. In 1981 I became a Chartered Physicist by virtue of being a Fellow of the Institute of Physics, implying that I had legal rights to practise as a physicist in the United Kingdom. However, I had by then realised that what I learned in physics in the 1940s was already outdated, and that my professional qualifications in physics were only of sentimental value to me.

Tun Abdul Razak succeeded Tunku Abdul Rahman as Prime Minister of Malaysia in 1970. I sat several times with him in a small group, which included the Deputy Prime Minster Tun Ismail and the Minister of Finance Tun Tan Siew Sin 陈修信 at the same table during lunchtime at Parliament House. I visited them at their residences on festive occasions. In 1971 Tun Razak declared an end to the emergency in the country and started to appoint his Cabinet. Tun Tan Siew Sin sent word to persuade me to join the Cabinet, saying that the Prime Minister had already agreed to make me a Minister by first appointing me a member of the Senate to bypass the normal route of going through the House of Representatives. I was not told which portfolio I would be given. In any case I declined politely, saying that I had neither interest nor expertise in politics, preferring to remain an academic. I was thinking that even if I chose to join the Cabinet, my position would not last longer than half a year. When I mentioned this to my Vice-Chancellor he agreed with my decision completely. He said that it would not be worthwhile for me to take up any of the portfolios except that of education, where I would have had something to offer, but then he doubted that I would ever be given that job. I also told my friend Hamdan about the offer and what I thought. I said that I would not be able to keep that job for more six months for the members of the party that I would have to join would get at me even before members of UMNO could have a chance to do so. Hamdan laughed and said that he was confident that I would be able to survive at least for a year. I could never imagine then that what I feared would happen to me did happen to none other than Tun Tan Siew Sin himself some four years later.

I was approached to consider a proposed position of a full-time Dean of Science at the University of Singapore, but I was reluctant to exchange research for pure administrative work. Rayson Huang offered me the post of

Professor and Head of the History Department at Nanyang University together with the Deanship of its Arts Faculty. I replied that for the time being I had to bring up the Tunku Abdul Rahman College, which was a responsibility handed over to me by him. I did not tell him another reason why I chose to stay in Kuala Lumpur at that time. I had a fight on my hands to defend the Department of Chinese Studies at the University of Malaya. By then I was one of the most senior professors in the university and enjoyed the trust of both the university authorities and the government. In the Department of Chinese Studies I had the full cooperation and friendship of my colleagues. I owed it to my university as well as to my colleagues not to desert them in their time of need.

Disputes between the history department and the Chinese department in a university over the teaching of Chinese history, or between a single discipline department and an area-studies department over the teaching of the discipline concerned relating to the area, are not uncommon. I was fortunate not to have encountered any such problem during the early period of my service at the University of Malaya. In fact, instead of disputes I had full cooperation from my fellow departmental heads. The heads of the history department and the Indian Studies department participated in giving students in my department the benefit of their expertise. The geography department did likewise with their visiting professor from Yale. Nobody had thought of the changes in the university atmosphere that came in the wake of the civil riots. Some members of the teaching staff, mainly the more junior ones, began to involve themselves actively in politics, including those within the university. The position of the deanship became furiously contested. In the Faculty of Arts there were the four so-called single-discipline departments of English, Geography, History and Islamic Studies, and the three so-called area-studies departments of Chinese Studies, Indian Studies and Malay Studies. The classification of Islamic Studies was less clear. By then there was also a language centre in the faculty. Then came a move to transfer all the "discipline elements" in the area-studies departments to the single-discipline departments, such that the Department of Chinese Studies, for example, would lose Chinese history to the History Department and Chinese language to the Language Centre. There was another proposal to have all courses conducted, and all dissertations for higher degrees to be written, in Malay, the national language. This would leave the Department of Chinese mainly with Chinese literature and teachers who were unable to teach in the official language medium, however good scholars they might be. The trouble was that these activists were riding with the tide making me uncertain whether they were implementing policies directed by higher authorities. I went to see Ungku Aziz, my Vice-Chancellor, asking whether there was such a direction from the government or from him and, if such be the case, I would not stay on to waste my time staging a futile fight against the activists knowing it would end in frustration. Ungku Aziz assured me that no direction had come from him nor the government as far as he knew. He said that he needed my support as a senior professor to deal with these activists together.

The Department of Chinese remained intact. But there were general university policies that it had to abide by. For example, all official correspondence and meetings within the university, and all courses opened to students outside the department had to be conducted in the national language. However, some courses in Chinese Studies could be taught in Chinese, and so were some courses in the English Department. Power struggles within the campus had changed the atmosphere in the university. Although members of the Department of Chinese did not involve themselves in this sort of activities, the department could not escape from being engulfed in the same atmosphere. Many issues that were not purely academic dominated some university meetings, consuming more time and energy than before. Fortunately, I managed to carry on with my research to maintain my morale. I wrote part of the article on ancient Chinese astronomical records for the *Physics Bulletin* mentioned above during university meetings while non-academic issues were being discussed. In the same year I received from the Cambridge University Press a copy of *Clerks and Craftsmen in China and the West* that incorporated works by Needham with the collaboration of Wang Ling, Lu Gwei-Djen and me.[48]

---

[48] See Needham, J. (with the collaboration of Wang, L., Lu, G.-D. & Ho, P. Y.), (1970), *Clerks and Craftsmen in China and the West* (Cambridge University Press).

In 1971 I was elected Fellow of the Institute of Physics. Needham would be visiting Kuala Lumpur in the later part of the year. To commemorate his visit, and with the help of the university librarian, Beda Lim, I wrote an article on a Song alchemist.[49] At the same time I collaborated with colleagues in the Department of Chinese Studies in research, partly to keep up their morale.[50] Needham came to Kuala Lumpur in September, accompanied by Dorothy Needham and Lu Gwei-Djen. They stayed with my family at our home in 170 Petaling Garden. I arranged for a public lecture to be given by him at the University of Malaya. The Vice-Chancellor, Royal Professor Ungku Aziz, introduced Needham to a packed-house audience as an intellectual giant. Needham also visited the Tunku Abdul Rahman College, besides the University of Malaya. There were two social functions arranged for my guests, one for Needham and one for Dorothy. I hosted a dinner at a Peking-style restaurant for my three visitors, inviting academics with Cambridge connections and scholars known to Needham as my guests. Among them were Justice Suffian, the Pro-Chancellor of the University of Malaya, and his wife, and Wolfgang Franke and his wife. Dr Chua Seong Siew, formerly of Gonville and Caius College, hosted a birthday party for Dorothy Needham at a Chinese restaurant in the Merlin Hotel. At home Lucy introduced the durian fruit to our guests. Needham asked for a photograph to be taken showing this exotic tropical fruit in front of him and my son Yik Hong. With three cars in the family, my chauffeur could take my guests around in a Mercedes Benz, while I went to work in a Vauxhall Cresta, and Lucy carried out her daily routine of sending and fetching our children to and from school and shopping in her Opel Kadett. Our guests seemed to enjoy their stay with us. During his visit Needham did not mention anything about my further involvement with *Science and Civilisation in China*.

---

[49] See Ho, P.Y. and Lim, B. (1972), "Ts'ui Fang, a forgotten 11th-Century Chinese Alchemist", *Japanese Studies in the History of Science*, 11:103–112.

[50] See, for example, Ho, P.Y. and Yu W.-L. (1974), "Physical Immortality in the Early 19th-Century Novel *Ching-hua-yuan*", *Oriens Extremus* (Hamburg), 21:3:3–51 and Ho, P.Y., Goh, T.C. and Parker, D. (1974), "Po Chu-i's Poems on Immortality", *Harvard Journal of Asiatic Studies*, 34:163–186.

After Needham left Kuala Lumpur I soon received news that the Australian National University in Canberra had nominated me as the Asian Fellow for the year 1972. The 1971 Asian Fellow was Li Chi, the Director of the Institute of History and Philology of the Academia Sinica, Taipei, and former teacher of my friends Cheng Te-k'un and Chang Kuang-chih. I did not have the pleasure of meeting Li Chi, but had corresponded with him officially to respond to his request to assess an article for publication. I informed Needham about my award. Hearing this, Needham sent me a copy of his gunpowder file and asked me to take this opportunity to visit Canberra to sit down together with Wang Ling to write the draft for the gunpowder epic section of *Science and Civilisation in China*.

In 1972 I made use of my sabbatical leave from the University of Malaya and special leave from the Tunku Abdul Rahman College to go to Australia for six months as Asian Fellow at the Australian National University. The Department of Far Eastern History in the School of Pacific Studies and the Department of Chinese in the Faculty of Arts jointly hosted my visit. Wang Gungwu was the departmental head of the former, while Liu Ts'un-yan was that of the latter. One of the first things I did was to see Wang Ling to tell him what Needham had asked me to do for him. I was pleased to hear that he had already done his job. I sent the good news to Needham and felt free to perform my function as Asian Fellow without carrying the burden imposed on me by Needham. I was not aware then that my report to Needham would give rise to some misunderstanding. The repercussion came only after my return to Kuala Lumpur.

As Asian Fellow my duty involved giving two public lectures, one each at the Department of Far Eastern History and the Department of Chinese. I had an office in each of these two departments and I arranged to be at the Department of Far Eastern History in the morning and at the Department of Chinese in the afternoon. The Australian National University would also pay for my travelling expenses to give lectures in universities throughout Australia and New Zealand. I gave lectures in several universities in Sydney and Melbourne, and attended a conference in Waikato, while giving lectures at a number of universities in the North Island of New Zealand. It was also in 1972 that I began publishing my research work in the Southern Hemisphere.[51]

I stayed at University House in Canberra. Wang Ling used to call for a game of either tennis or table-tennis. I heard that he was quite a good table-tennis player, but unfortunately he had a heart condition that deterred us from having a serious game together. Some members of University House often met for a game of chess. That was when the Fisher versus Spassky chess matches were reported in the news. The top player at University House then was a Polish astronomer. I had a couple of games with him, all ending in a draw. We attracted a small crowd of watchers, some of whom dubbed our games as mini Fisher versus Spassky meetings. I also had several games of chess with Professor K. Mahler, F.R.S., "who not only raised many queries in the mathematical section, but even undertook special research where obscurities needed to be illuminated" in Needham's *Science and Civilisation in China*, volume 3.[52] Mahler was also a friend of Oppenheim, but I did not ask whether they played chess together. Unfortunately, Mahler was then in his eighties so mental fatigue prevented him from showing his usual best. Another member of University House I often met, but not for a game of chess, was Y.S. Chan 陈炎生, the in-coming East Asian Librarian of the University Library. He later became very helpful when I needed reference material in his library.

Life in the tranquil surroundings of University House provided me an opportunity for self-assessment. In the past two decades I had devoted myself to work in the university and off-and-on to Needham's requests. I was quite contented with my lifestyle, without thinking about what would come after the next one or two decades. I thought that I was still young and did not ponder over the future of my children's education. I was then aged 46 and suddenly realised that I would have to retire from my service at the Univer-

---

[51] See Ho, P.Y., Goh, T.C. and Lim, B. (1972), *Lu Yu, the Poet-Alchemist*, Australian National University *Asian Studies Occasional Paper*, No. 13 (Canberra); Ho, P. Y. (1972), "Early Chinese Science", *Hemisphere*, 16:10–14 and Ho, P.Y. (1973), "Magic Squares in East and West", *Papers on Far Eastern History*, 8:115–141.

[52] See page xlvi.

sity in nine years when I reached 55. In two to three years' time my son would reach university age, to be followed a year later by his two twin sisters. There might be some possibility for one or more of them to win a scholarship, but not very likely, if they succeeded in getting into a local university in Malaysia. There was also some risk that they might not be able to study subjects of their choice. There might be the scenario of having to send three of them overseas for university education. Then about the time I reached the age of 51 I would have to do the same for my third daughter and another three years later for my fourth daughter. All these would mean that I would probably have to support at least one daughter studying overseas after my retirement.[53] I had already paid up the mortgage for my house and held a small portfolio of company shares and fixed term deposits in the bank. Supporting my son to study overseas would not create too much of a problem, but supporting three children together would be quite a stretch on my financial resources. Problems would probably arise when it came to supporting my last two daughters before and after my retirement. One common practice in my part of the world was to make the older children sign a bond to support their younger brothers or sisters after their own graduation, but I would hate to impose any bondage on my children and would not wish to rely on them financially even in my own old age.

While I was pondering over the future I heard about the establishment of two new universities in Australia, namely Griffith University in Queensland and Murdoch University in Western Australia. Griffith University was to establish a School of Modern Asian Studies and invited applicants to fill a position of Foundation Professor in the School. There would be two Foundation Professors in the School; one of them would also be appointed Chairman for a term of five years. The position of Foundation Professor was tenable up to the age of 65. I submitted my application for the post of Foundation Professor just before I left Canberra for Kuala Lumpur, quoting Needham's name as one of my referees. Curiously enough, that happened to be my first and only time in life to apply for a job.

---

[53] It turned out eventually that four of my children chose to study medicine.

I returned to Kuala Lumpur in July 1972. Things were quite normal in the Department of Chinese Studies at the University of Malaya. I noticed, however, some subtle change in the attitude of certain members in the College towards me. The Chairman of the governing body of the College, Tan Sri Khaw Kai Boh 许启谟, passed away in England during my absence in Australia. A number of members of the governing body, who were also active members of the MCA political party, wished to replace me by appointing a new full-time Principal. The candidate they proposed happened to be a friend of mine looking for a job. I would have gladly agreed to pass over my job at the Rahman College to their candidate, but I remained silent while things were kept secret from me.

In the month of September I received an invitation to go to Brisbane for an interview with the appointment committee of Griffith University. The interview lasted a whole afternoon. Someone even referred to Lucy's expertise in cooking which was mentioned in Needham's referee report. Towards the end of the interview the Vice-Chancellor said that I could expect an offer for the appointment of Foundation Professor and first Chairman of the School of Modern Asian Studies, and asked me how soon I could be on board. I replied that I needed to give six months' notice to resign from my post at the University of Malaya.

After my return to Kuala Lumpur I received an invitation from the new Chairman of the governing body of Abdul Rahman College to a meeting, at which he asked me to apply for the post of Principal in the College. When I expressed my disinterest in applying, I noticed a sign of relief among those present. It appeared to me that the President of their political party, Tun Tan Siew Sin, was likewise kept in the dark over the intended appointment of a new Principal. I was thinking that if I had joined the government as a cabinet minister, the same group of people could have done something similar to me. Luckily on this particular occasion I could anticipate that sort of action and took things with a smile.

The problem of Wang Ling surfaced upon my return to Kuala Lumpur. Upon hearing from me that Wang Ling had said that he had already done his job, Needham wrote to Wang Ling to congratulate him and asked him to deliver his draft on the gunpowder epic. Wang Ling replied that he was ready to work with Needham in Canberra. In fact, at the request of Wang Ling, Wang Gungwu had arranged for an invitation to be extended to Needham to visit the Australian National University as well as an honorary doctoral degree to be conferred on him during his visit. From Wang Ling's point of view, he did not mislead anyone when he said that he had already done his job. Since the very beginning his collaboration with Needham consisted of searching for material, writing cards and preparing notes. Actual writing of any part of *Science and Civilisation in China* was Needham's responsibility. After all, even Lu Gwei-Djen herself did not do any writing. Of course on this matter Wang Ling was not in the same position as Lu Gwei-Djen who was able to say candidly to me the following words about Needham: "He can write better. Why should I write?" The whole picture became clear to me. Wang Ling was doing his very best to get Needham to Australia to write the gunpowder epic. I could not blame him for rejecting me as a substitute for Needham. On the other hand, however tempting it was to visit the Australian National University, it was not possible for Needham to make the trip. Besides the fact that his term as Master of Caius was ending soon, working away from his cards and files to use unfamiliar substitutes prepared by Wang Ling was not a good proposition. Furthermore, Needham could not find any excuse for working with Wang Ling on the gunpowder epic instead of Lu Gwei-Djen to write the medical volume, which was why he picked me as the solution.

Letters came from Needham and Peter Burbidge, the project coordinator, asking me to take over the writing of the gunpowder epic from Wang Ling. That put me in a difficult position. Those who made demands on me seemed to ignore my future position in Australia, both regarding Griffith University and Wang Ling. Wang Ling was a delicate issue. Although Australia is a big country, the circle of scholars in Chinese studies was then quite small. Everyone within the circle knew Wang Ling and about his collaboration with Needham. To take over the writing of the gunpowder epic section would not only annoy a friend but might also be misinterpreted by some as robbing something from a friend. I explained my position to Peter Burbidge, saying that I would take over drafting the gunpowder epic section provided Wang Ling's name would not be removed, and also mentioned his early interest in gunpowder when he and Needham first met in wartime China. Lu Gwei-Djen later told me that my letter caused some consternation at a meeting in Cambridge until she pointed out that all I wanted was to retain Wang Ling's name, something that Needham readily agreed to. The gods over there did not seem to understand the problems of the outside mortals.

In December 1972 I received the official letter of appointment from Griffith and went straight to inform my Vice-Chancellor, Royal Professor Ungku Aziz. He said that he knew he would not be able to retain me this time. The next day he received a letter from the Vice-Chancellor of Griffith University asking him to release me as early as possible as a favour to help a sister Commonwealth university in its establishment. My new appointment was announced simultaneously in Australian and Malaysian newspapers. Ungku Aziz, with the consent of the university council, reduced my contractual period of notice of resignation from six to three months. Tun Tan Siew Sin congratulated me on my new appointment and thanked me for my past service at the Tunku Abdul Rahman College. (See Figure 11.) It was unthinkable then that, two years later I had to write him a letter of consolation on his retirement, which resulted from the similar cause that I had feared would happen to me had I accepted his earlier offer to join him in the Malaysian Cabinet.

I was fully occupied during the first three months of 1973 preparing to leave for Australia. I had to arrange for my books and household belongings to be packed and shipped to Australia, sell my house at 170 Petaling Garden and dispose of some of my stocks and shares. As my mother was still in Malaysia I left some assets behind in Kuala Lumpur under the custody of a bank for her support and for contingency. I had to arrange for my five children to leave their schools in Kuala Lumpur and find new ones for them in Brisbane. In addition we had to attend farewell parties from friends and colleagues.

At the end of March I set off for Australia with a heavy heart, leaving behind my aged mother, my brothers and sisters along with my many colleagues and friends. Consoling myself that I would henceforth work to bring Australia and Asia culturally closer together, I plunged into the uncharted

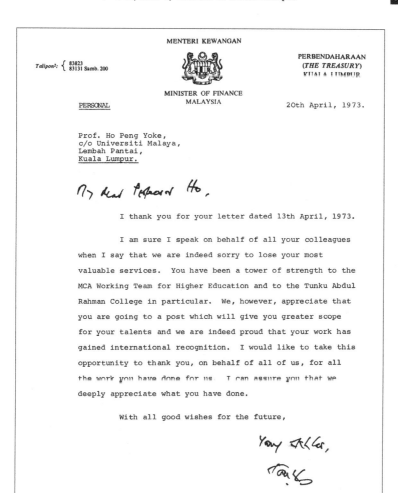

Figure 11 Letter from Tun Tan Siew Sin

waters of a new Australian university that aimed to be different from the conventional. On the mundane side, I knew that I had extended my retirement age by ten years. My children not only received free university education in the subjects of their choice, but were also able to graduate one or two years earlier than their contemporaries back in Malaysia. I had no idea then that the move I made to go to Australia would one day enable me to have sufficient means to take up a non-stipendiary post in Cambridge.

4

A New Australian University

I arrived in Brisbane with my family after mid April 1973 to take up my new appointments at Griffith University. The university helped me to lease a house nearby in the suburb of St. Lucia for three months. Four of our children gained admission into Toowong State High School, while our youngest, Sook Pin went to the Ironside Primary School. On 8 July we moved into a house that I bought at 8 Holdway Street in the suburb of Kenmore. Sook Pin changed to the Kenmore State Primary School which was nearer home, while her brother and sisters continued to attend the same school. Lucy sent and fetched our children to and from school in our own car, while I used my official car to travel on official duty and between home and the university. The university provided an official car to each executive officer in lieu of allowances.

There was no building and not a single brick was seen at the site for the new university in the suburb of Nathan. There were neither students nor library books to speak of, only trees. The university rented the entire sixth floor of Sherwood House at 37 Sherwood Road, in the suburb of Toowong, which was nearer to the city centre and the University of Queensland, as its temporary administrative office. The Vice-Chancellor, Professor F.J. Willett, was the chief executive officer. Under him were eight Divisions, each with its own executive officer. These executive officers consisted of the Business Manager, Registrar, University Librarian, University Site Planner and the four Chairmen of Schools, namely the School of Australian Environmental Studies, School of Humanities, School of Modern Asian Studies and the

School of Science. In the absence of the Vice-Chancellor, one of the four School Chairmen, determined by the seniority of coming on board, would act in his place. Fortunately I was third in priority, so that I only had to act as Vice-Chancellor for a period of just more than a week one year before my term as Chairman of School expired. The rather unusual titles, such as Business Manager in place of Bursar, School in place of Faculty or College, Chairman in place of Dean came about as the university was seeking something different from the conventional.[54] A case in point was the Chancellor, Sir Theodore Bray, who also played the part of Chairman of the University Council and participated actively in administrative affairs.

As the first and most senior member of the School of Modern Asian Studies, I played an active part in the appointment exercises of not only academic and non-academic staff in the School of Modern Asian Studies, but also those in the other seven Divisions. I attended meeting after meeting almost every day, discussing the choice between the three-academic term and the two-semester systems, the setting up of university regulations, the drawing of blueprints for the new university buildings, the employment of teaching and administrative staff members, drawing up budgets for the university, and so forth. For example, after more than six months of discussion the two-semester system was decided on. Then came a newly appointed Foundation Professor in another school who happened to have made my acquaintance when I was Asian Fellow in Canberra in 1972. He was a strong believer of the three-academic term system and asked for my help to reopen the issue. I replied that I would not oppose him, but would not wish to waste more time after having spent over six months on the issue. I suggested to him to speak to the other colleagues, and should he succeed in

---

[54] After the retirement of the first Vice-Chancellor, his successor changed the title of Chairman to Dean of a Division. However, by then some Schools had lost their former status as Divisions. The School of Modern Asian Studies became part of another Division and still retained a Chairman. As I was Chairman of a Division during my term of office, the new Vice-Chancellor, Professor Roy Webb, referred to me as the first Dean of the School of Modern Asian Studies. By now the School of Modern Asian Studies has lost its name in Griffith University.

convincing the majority I would readily abide by it. He went to speak to the Vice-Chancellor and his colleagues, but could not get any further. The idea of having the two-semester system originated from the Vice-Chancellor himself. To me what was more important than the system were the quality of the teachers and the students, adequate library books and laboratory equipment, followed by student amenities. I was quite happy to abide by the decision of the majority regarding this issue.

In order to attract school leavers to apply for entry into this new university, all members of the teaching staff were enlisted to visit secondary schools within and outside Brisbane and talk about Griffith University and its four Schools before giving some emphasis on the School the speaker came from. The difference between the new university and the existing University of Queensland would be pointed out in the talk. For my part, I talked mainly about learning Chinese and the importance of understanding the culture of the Asian people. I tried to arouse the interest of my audience and to speak in simple terms. For example, I once wrote the Chinese word *"women"* on the white board and remarked that all Chinese referred to themselves with that term in Chinese. That woke up and amused the audience. I then proceeded to break up the word into two parts, one referring to the first person and the other giving the plural number. I congratulated my audience that they had already learnt to say in Chinese the two words "I" and "we" quite effortlessly. A member of the audience once asked whether Chinese could be a precise language, since it had no number, no tense and no conjugation. Knowing that at least half the audience had taken French in school I asked whether they agreed that French was a precise language. When they agreed that it was, remembering the little French I learned at school, I wrote on the white board the sentence "I follow an ass" and asked if anyone would volunteer to translate it into French and say it aloud. When no volunteer came forward I explained that I was not trying to show that French was imprecise, but rather to demonstrate its subtlety. Chinese could be written precisely if required by adding the number, the time, the gender, etc. On the issue of understanding the culture of the Asian people, I once gave an example of shopping in an Indian shop, where one would try his or her skill in bargaining. I told my audience that when the shopkeeper shook his head one should stop bargaining instead of continuing to raise the offer, at which he would shake his head even more. I then explained that when the shopkeeper first shook his head, he had already agreed on the offer, because in this case shaking should be interpreted as nodding his head.

Activities such as these had no bearing on scholarship; they are just mentioned here in passing. I sometimes had the feeling that I had already become an administrator if not a salesman. To get to know the local Chinese community I joined its Chinese Club and attended its dinner functions, sometimes inviting my Vice-Chancellor and his wife as guests. This was to meet the hope of my Vice-Chancellor to gain support from the Chinese community.

I had little time for sports and recreational activities. I bought a new tennis racket for my son Yik Hong, who took me to the University of Queensland's tennis court to try out his new acquisition. Although the racket was bigger than what I was used to, I missed the ball altogether as if it had gone through the racket. I put the blame on the fact that I was not used to the racket, perhaps without realising that some defect in my vision was developing. That was the last time I stepped on a tennis court. I installed a table tennis-table in my home; my game had deteriorated, which could partly be due to my vision. In my student days I did not understand the game of cricket and thought of it as a waste of time. In my undergraduate days in Singapore, I once watched a classmate, who was the captain of the college cricket team, getting four runs when the ball hit the edge of his bat. I thought it was a deliberate skillfully executed stroke that eluded all the fielders behind him. When I joined Griffith University I was asked to be one of the umpires in a friendly staff match. The other umpire was the Vice-Chancellor. The players did not accept my excuse of ignorance of the rules of the game. They said that they would abide by my decision. As soon as the bowler appealed for leg-before-wickets I raised my hand and did it consistently for both sides. The game was over within two hours; I was never approached again to be a cricket umpire. Nevertheless, the popularity of cricket in Australia made me learn a little more about the game to the point that I could follow a match shown on television.

A friend and former university colleague of mine in Singapore, Dr Damodar Singhal was a Professor of History at the University of Queensland. Lest I forgot my position as a scholar and teacher, he invited me to offer a course on the history of science at his department.[55] I agreed to teach the course on an honorary basis, without any form of remuneration, so the Faculty of Arts treated me as an honorary faculty member. I said that I was only too happy to do it that way to show my appreciation of my children receiving free university education in that university. The course on the history of science turned out to be very popular. Even one of the Deputy Vice-Chancellors, Professor Edwin Webb of the Biochemistry Department, enrolled as a part-time Arts student and signed up for the course. After adding a degree in the humanities to his doctoral degree in science, he left the University of Queensland to take up a new post in Sydney as Vice-Chancellor of Macquarie University.

The History Department of the University of Queensland also had a tutor, who enrolled as a Ph.D. candidate working on ancient Japanese history, but it could not find him a supervisor. It sought my help saying that the tutor was a very good teacher, but would be out of employment at the university if he could not get a doctoral degree. Remembering what Oppenheim did for me, I consented. The thesis concerned Korea in the early history of Japan, which was then a sensitive issue. I had to be careful in finding the two external examiners for his thesis. Eventually, I found him two native Japanese external examiners, one from the University of Chicago and the other from Kyoto University. The candidate himself was a native Japanese. He got his Ph.D. degree, but left the University of Queensland for a lecturer's post in a university in New Zealand. Besides helping the University of Queensland, I also served on the Board of Advanced Education of the State of Queensland.

---

[55] One of the lectures appeared in a book dedicated to the memory of Professor Damodar Singhal. See Ho, P.Y. (1992), "History of Indian Science", in Arvind Sharma, ed. *Perspective on History and Culture* (Delhi) pp. 219–226.

As a result of the keen interest of the Vice-Chancellor and the Chancellor of Griffith University in modern China, I had the opportunity to make my first visit to Mainland China. Being then a Malaysian citizen, I had to seek permission from the Deputy Prime Minister and Minister for Home Affairs of Malaysia, Tun Dr Ismail al-Haj, as Malaysian citizens in those days were restricted from visiting China. The Malaysian Government readily granted me permission. In November 1973 the President of the Association for the Friendly Relations with Foreign Countries, Zhu Qizhen 朱启祯, invited a team of five from Griffith University in Queensland, Australia, to visit China. The team consisted of the Vice-Chancellor of the University Professor F.J. Willett and his wife, the Chancellor of the University Sir Theodore Bray and his wife, and I as the Head of the School of Modern Asian Studies. This invitation came after Needham had spoken to the Chinese ambassador in London about our intention to visit China.

We visited the five cities of Guangzhou, Beijing, Nanjing, Hangzhou and Shanghai, and the four universities of Zhongshan, Beijing, Qinghua and Fudan. At the Qinghua University we were led into its computer laboratory for a demonstration of printing that showed five Chinese characters reading "Long Live Chairman Mao" — quite a feat in yesteryears. We were among the early groups of academics visiting China after President Nixon. At the Dongfang Hotel in Guangzhou I met Arthur Wright, who was then leading a team of American archaeologists on a study tour. I caught up with him again later at the Beijing Hotel in Beijing. I met Cao Tianqin 曹天钦 for the first time at the Institute of Biochemistry in Shanghai.[56] Cao's name was rendered by Needham as Tshao Tian-Chhin and was included in one of our joint publications on Chinese alchemy. Cao played an important

---

[56] Our second meeting was in Brisbane two years later when he paid an official visit to Australia and became my guest at a dinner party. Our third meeting was in 1978 in Shanghai when he and his wife (Xie Xide 谢希德) were our guests at dinner when my wife and I stayed at the Jingan Hotel. Our last meeting was in Beijing and Shanghai in 1984. In Shanghai my wife and I, together with Needham and Huang Hsing-tsung, were invited to dinner by Cao Tianqin and Xie Xide at their residence. Xie Xide was then the President of Fudan University.

role in the synthesis of bovine insulin during the 1960s. He showed me a model of its molecule displayed in his Institute, besides introducing me to the Director of his Institute, Dr Wang Yinglai 王应睐, who was one of the three Chinese students who went to Cambridge in 1937 to do research on Biochemistry.[57]

At Beijing we met Dr Zhou Peiyuan 周培源, famous Chinese physicist and President of Beijing University, the well-known Chinese mathematician Hua Luogeng 华罗庚 and Professor Zhou Yiliang 周一良 of Beijing University at a dinner party hosted by the Australian ambassador, Dr Stephen Fitzgerald, at the Peking Duck Restaurant. Although it was the first time that I met these three scholars we spoke like old acquaintances, because we all had Needham as a mutual friend. Needham and Lu Gwei-Djen visited China in 1972, after which they heard that the latter was hospitalised with a lung problem. They received no further news, and some feared that she did not survive. Hua Luogeng asked me to confirm this, but I had no recent news about her either. After my return to Australia I wrote to Needham, discreetly enquiring about the health of both Dorothy Needham and Lu Gwei-Djen and found that both of them were well. I wrote to inform my new friends in China.

We left China via Guangzhou, where our host gave us a farewell dinner at the Beiyuan 北园 Restaurant. There were then only three famous restaurants in Guangzhou, namely the Beiyuan, the Panxi 泮溪 and the Nanyuan 南园. We took a train to Hong Kong where I had to remain behind, while the rest of the partly returned to Brisbane. At the request of Rayson Huang I delivered a public lecture at the University of Hong Kong on the history of Chinese astronomy. When I assumed duty at Griffith University, Rayson Huang had already left his post as Vice-Chancellor of Nanyang University in Singapore to become the Vice-Chancellor of the University of

---

[57] The three of them were Lu Gwei-Djen, Wang Yinglai and Shen Shih-chang 沈诗章. Needham was Shen's supervisor. Shen took up an academic career in the US and did not participate in Needham's *Science and Civilisation in China* project. I have not met him either.

Hong Kong.[58] As noted earlier, he was the person who introduced me to Needham and played a role in my appointment to the Chair of Chinese Studies at the University of Malaya. I was only too happy to oblige when he asked me to give a public lecture at his university.

I returned to Brisbane by way of Kuala Lumpur in order to visit my mother, brothers and sisters in Malaysia. I happened to sit next to a friend, Hamzah Majeed, whilst travelling on the same plane to Kuala Lumpur. He was the Personal Secretary to the Prime Minister of Malaysia, Tun Abdul Razak. He said that I should go and see the Prime Minister, as he would be interested to hear about my visit to China. I told him that with the flu that I had contacted, it would not be a good idea for the Prime Minister to see me, but I would write instead. I stayed at the Merlin Hotel in Kuala Lumpur and asked the hotel secretary to help me type a letter to Tun Abdul Razak telling him about my visit to China. I wrote to Tun Razak again shortly after my return to Australia and sent him a copy of the lecture on Chinese education to be delivered at the University of Adelaide. See Figure 12.

Both the Vice-Chancellor and the Chancellor, together with their wives, were very friendly to Lucy and me. Having read Needham's comment about Lucy's culinary skills, they were always looking forward to be invited to dinner at our home. As a result we invited guests regularly to our home for dinner. Our guests included not only members of the university staff, but also others, such as the Japanese Consul-General in Brisbane and his wife, Sir Zelman Cowan, Wang Gungwu, Hamdan, and so on. My Chancellor, who was a retired journalist, was impressed by my knowledge of languages when I went to Brisbane for an interview. He was greatly impressed by my performance during our China trip. He overheard me speaking Japanese to a Japanese passenger travelling on the same plane to Hong Kong. He noted

---

[58] When Cheng Te-k'un was working in the University of Malaya, he referred to Rayson Huang, Wang Gungwu, himself and me as the four most capable scholars of Chinese descent in the country. Interestingly, the four of us eventually all landed up in Hong Kong. Rayson Huang was the first, followed by Cheng Te-k'un, then me and Wang Gungwu, in sequence of time.

```
Nathan.  Queensland.  4111.  Australia.



20th February, 1974.



Y.A.B. Tun Datuk Haji Abdul Razak
    bin Datuk Hussein, S.M.N., D.P.M.P.,
Prime Minister of Malaysia,
Jalan Datuk Onn,
Kuala Lumpur,
MALAYSIA.

Dear Tun,

I am forwarding to you a copy of the lecture on 'Higher Education
in China' which I shall deliver next month at the Seminar on Asian-
Australian Educational and Cultural Relations in the University of
Adelaide.  The lecture reflects some of the observations I made
during my brief visit to China last November.

With best wishes.

                                   Yours sincerely,



                                   Hu Peng Yoke
```

Figure 12 Letter to Tun Abdul Razak

the warm reception we received from my friend Rayson Huang, the Vice-Chancellor of the University of Hong Kong. When we checked into our hotel in Hong Kong he saw me greeting Arthur Wright of Yale University. In China he noticed the friendliness of some of the top scholars in China towards me. He concluded that I was an internationally known scholar with the rare ability of knowing all the Asian languages taught in Griffith University. That, however, did not do me much good. I had never taken advantage of his kindness to seek any assistance from him in university affairs. On the other hand, being an elderly good-natured gentleman, he sometimes complimented me in front of my peers. That did not make my work easier, especially when I had to try to escape from some committee work in order to attend to Needham's gunpowder epic.

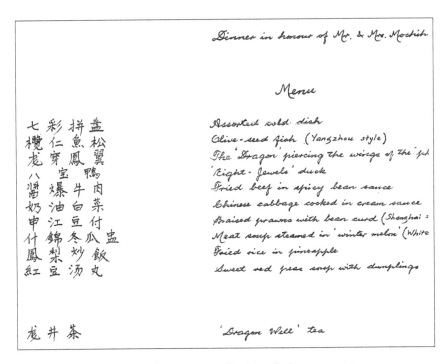

Figure 13 A typical menu prepared by Lucy for home entertainment

National and International Activities

In March 1974 I attended the centenary celebration international conference on higher education at the University of Adelaide and presented a paper on higher education in China.[59] Rayson Huang, the new Vice-Chancellor of the University of Hong Kong was also an invited participant at the conference. We exchanged news about our families and talked about Needham. He also told me about his son, Christopher gaining an exhibition from Oxford to

---

[59] See "Higher Education in China", *Proceedings of the University of Adelaide Centenary Celebrations: Seminar on Asian-Australian Educational and Cultural Relations, Past, Present and Future* (1974), pp. 101–109.

read medicine.[60] We had a warm reception in more ways than one during a time when hot wind from the desert blew south towards the city. That year the University of Hong Kong conferred an Honorary Doctor of Letters Degree on Needham. Some time in 1974 Griffith University moved from Toowong to another commercial building near Evans Road in the suburb of Salisbury which was closer to the Nathan university site. By then the construction of the university buildings was already underway. We were there for quite a few months, but it seemed that many of us had lost recollection that we once worked at that site.

In August 1974 I went to Tokyo and Kyoto to attend the 14th International Congress of History of Science. I arrived in Tokyo on 15 August, three days before the opening of the conference and checked in at the Shimbashi Daiichi Hotel. On 16 August I moved to the Norin Nenkin Kaikan Hotel as a guest of members of the *Ippōkai*. After joining my hosts on a sightseeing bus tour around Tokyo, I attended a welcome dinner reception at the hotel. The next day my hosts took me on their annual picnic tour to the hot springs of Kinukawa and to Nikko, staying overnight in the Hotel Takahara in Kinukawa. On my return to Tokyo I checked into the Shimbashi Daiichi Hotel, where the Soviet delegates to the conference were also staying. At the reception party many of them gathered around me to ask about China; somehow they had heard that I was in China the previous year. I told them frankly that I had a most enjoyable visit and saw only what I was shown. I then said I should learn from them about China, since they should know more as a neighbour than one living far away in the Southern Hemisphere. I saw Needham in the company of someone, who, at first glance from a distance, appeared to me as Lu Gwei-Djen, but somewhat slimmer and more energetic. I thought that the slimness was due to her long illness. However, on approaching the two, I found Needham with somebody else. Needham introduced her to me as Shih Hsio-yen 时学颜, Curator of Museum in Toronto. They stayed at the Shimbashi Daiichi Hotel for

---

[60] Christopher Huang is now Reader in Physiology at Cambridge University and Fellow of New Hall.

the first night, but moved to the Fairmont Hotel on the second day. Shih Hsio-yen told me that Needham could not find a desk large enough for him to do his work so he decided to move out to a bigger hotel. I did not see much of them at the conference and neither did Needham ask me about my progress on the gunpowder epic. Many years later I heard that Lu Gwei-Djen was unable to accompany Needham on that occasion because she was unwell. On 24 August I went to Kyoto for the second part of the conference and stayed in the Kyoto Hotel. I chose to stay there because it was Kiyosi Yabuuti's favourite hotel and because of its proximity to the Kyoto University. At the conference I chaired its section 9 at the afternoon session of 21 August and I also read a paper on Chinese alchemical and medical prescriptions.[61] On 28 August I visited Osaka, staying at the Osaka Grand Hotel. From Osaka I returned to Brisbane via Hong Kong, Kuala Lumpur and Singapore to attend to some official business, while visiting my relatives.

In 1975 Griffith University took in its first batch of undergraduates. There were only first year students, but preparations had to be made for the second year teaching. The School had to appoint more teaching staff. People then had very little knowledge of East Asia. I remember on one occasion we interviewed an American-trained Japanese scholar for a lecturer post and invited her out to a restaurant for dinner afterwards. At the dinner we overheard two other guests sitting nearby talking about Japan. One of them said aloud that Japan had nothing to offer, except perhaps the art of flower arrangement. My Japanese guest was taken aback, suspecting some anti-racial sentiment. With a smile I looked at her and said, "Did you hear that? That shows how much you are needed here to tell them about Japan." She subsequently accepted our offer of appointment.

Three scholars of international fame visited me at the School of Modern Asian Studies. They were Cheng Te-k'un, Wolfgang Franke and Chiang Yee,

---

[61]"Chinese Alchemical and Medical Prescriptions — a Preliminary Study", *14th International Congress of the History of Science Proceedings No. 3* (1974), pp. 295–298.

the "Silent Traveller" from Columbia University.[62] They all talked to the students, while Chiang Yee also presented a painting of the panda to the School. To promote bilateral relations between Griffith University and other universities in Southeast Asia, I invited two Malaysian scholars to visit and give lectures in Griffith University and other establishments of higher education nearby. My friend Tan Sri Raja Mohar bin Raja Badiozaman, the Chairman of the Malaysian Airlines System, had kindly provided First Class airtickets to make such arrangements possible. The Professor of History of the University of Malaya, Zainal Abidin bin Abdul Wahid, was our first guest under this arrangement. Our next guest was Hamdan, the Vice-Chancellor Designate of the *Universiti Sains Malaysia* in Penang.

I left Brisbane on 21 November 1975 for Tokyo to accept an invitation from Keio University as the Leverhulme Visiting Professor at its Institute of Culture and Linguistic Research. On my way I stopped over in Hong Kong to help the editorial board of the *Journal of Asian Studies* to assess an article on the history of Chinese astronomy that was submitted to that journal for publication. Mr Ichiro Ishitani 石谷一郎, Chief Administrator of the International Centre of Keio University, and Mr Akira Fujii 藤井明 of the Daito Bunka University met me at the Haneda International Airport. They took me to a university-rented apartment at Futaba Mansion, 3-4-18 Mita, Minato-ku, Tokyo. I arrived in Japan during the period of a railway workers' strike. Fortunately Futaba Mansion was only about five minutes walk from the International Centre, where I had the use of an office, so I managed to go to my office as well as the library throughout the duration of the strike. My friend Shigeru Nakayama guided me to the National Diet Library.

My appointment in Keio University carried the official title Visiting Researcher. A Researcher in a university or research institution in Japan and China is equivalent to a Research Professor in some other countries and is

---

[62] See page 79. Chiang Yee served as a magistrate in China. He later travelled outside China while speaking little English. After acquiring a knowledge of the language, he wrote several books on his travels. That made him famous and gave him the name "Silent Traveller". His expertise in painting the panda and his calligraphy also earned him fame. He became a professor at Columbia University.

also known as a Professor. However the appointment carries no teaching and administrative duties, except under special arrangements as in the case of the Director. Taking advantage of my freedom, I wasted no time in working on the gunpowder epic for Needham's *Science and Civilisation in China*. A crucial text for my research was the *Huolongjing* 火龙经 (Fire Dragon Manual). I had four different versions of the text on microfilm, sent to me by Needham, which were taken from his personal collection in Cambridge. I came to know that there was another version of the text under the title *Wubei huolongjing* 武备火龙经 (Fire Dragon Manual for Military Armament), preserved in the library of the Boei Daigakko 防卫大学校 Military Academy in Mabori Kaigan. Shigeru Nakayama went with me by train to the Academy and managed to obtain a Xerox copy of the text. The book previously belonged to Seiho Arima 有马成甫, an expert on the history of gunpowder weaponry. There was a little amusing story on our way back to Tokyo. We stopped over in Yokohama for lunch and went to a Chinese restaurant in Chinatown, called the Yangzhou Restaurant. The chef told us that he was from Yangzhou. As Yangzhou fried rice was famous among overseas Chinese as well as in Hong Kong, I ordered Yangzhou fried rice. The chef looked perplexed and said that there was no such thing in Yangzhou cuisine. We ordered something else for our meal. On 17 December I gave a joint-staff seminar on the book for the Department of International Relations and the Department of History of Science at the Komaba campus of Tokyo University. I spent most of my time at the National Diet Library; it was also my friend Shigeru Nakayama who first guided me to this magnificent library.

I visited the Daito Bunka University and met Professor Goro Yoshigawa 吉川五郎, Professor of Chinese and Dean of the Faculty of Foreign Languages. Together with several members of his staff, including Akira Fujii, we discussed the exchange of staff and students between that university and my home university in Australia. The Dean was particularly happy to be able to have the whole proceedings conducted in Japanese. A professor in Chinese in Japan did not necessarily speak the modern Chinese language; he could read Chinese in some old Chinese tongues that came to be known as *onyomi* 音读.

I called at the Hongo campus of Tokyo University to see Professor Hiroharu Seki 関寛治 of the Research Institute of Oriental Culture. He was then involved with the Institute of Peace Studies in Hiroshima. I also went to Waseda University to see Professor Kinichiro Toba 鸟羽钦一郎. I knew both Seki and Toba when they were visiting professors at the University of Malaya. Another former visitor to Kuala Lumpur whom I met in Tokyo was Professor Fumiko Koide 小出词子, Japanese language expert at the International Christian University. I also hosted a dinner to welcome a new appointee who was joining the School of Modern Asian Studies in my university and helped the Language Centre of my university to interview applicants for the post of Japanese language lecturer.

I paid a courtesy call at the Japan Foundation to express my gratitude for the gift of books and the posting of a Japanese visiting professor to Griffith University. I also spoke to the Foundation to invite the Vice-Chancellor of Griffith University and his wife to visit Japan. Originally, the Japanese Consulate-General in Brisbane had written to the Japan Foundation to invite me to Japan, but I suggested that my Vice-Chancellor and his wife be invited as well. The Japan Foundation asked whether a visit by me was more important than a visit by my Vice-Chancellor and his wife. I answered that I had visited Japan many times, whereas my Vice-Chancellor had never been to Japan before. I added that because of the special interest of that university in Japan, besides China and Southeast Asia, it was important for my Vice-Chancellor to get to know more about Japan, more so becasue of his past experiences as a pilot in the Pacific War.[63] At the request of my Vice-Chancellor I also made a feasibility study of establishing a campus in Japan for Australian students learning Japanese to get more exposure to speaking that language, and of seeking help to build a Japanese garden within the Nathan campus of the university. I went to Okayama to obtain free expertise advice from my friend Saburo Kodera on finding a suitable site and the

---

[63] My Vice-Chancellor and his wife subsequently received an invitation from the Japan Foundation in 1976. From all accounts they enjoyed their trip very much. I suggested to them a list of the universities they should visit and informed the universities concerned in advance.

cost involved. During my visit to Okayama, Saburo Kodera took me to see Okayama University. The Aisawa Company, of which my friend was a member of the management, was a major benefactor of that university. There I met Professor Sennosuke Fukuda, the Dean of Students and an expert in Chinese classics. At that time the university was celebrating its library's acquisition of its one-millionth copy of books. I wrote a congratulatory message in the Japanese manner for the occasion. My friend also showed me a Japanese traditional Chinese school with a Confucius temple. I then went to Hiroshima and Miyajima for sightseeing. In Osaka I was the guest of a priest, whose sect was famous for its Japanese garden. My host put forward the idea that his sect would build a Japanese garden for Griffith University subject to a personal condition that I could not accept.[64]

In January 1976 Lucy visited me together with her elder sister Fanny and her younger sister Alice. I took them on a tour to Kyoto and Kyushu. On reaching Okayama we met my friend Saburo Kodera, who joined us all the way to Fukuoka. A number of my Japanese friends were there to host a dinner to welcome us. My friend Shoji Katano 片野昭二 was then working in Fukuoka as Head of the Department of Administration Control. He arranged for me to meet the Lord Mayor of Fukuoka, Alderman Shinto, on 16 January. The Mayor told me that he had discussed with his counterpart in Brisbane about forming a link between Fukuoka and Brisbane as sister cities, but the final decision had to await the election of the next Mayor and the City Council in Brisbane in March 1977.[65] My friend Katano accompanied Lucy and me, together with Lucy's two sisters, to Nagasaki and Kumamoto, and across the volcanic area of Mount Aso to the hot spring resort of Beppu before we returned to Hakata in Fukuoka to board a *Shinkansen* bullet train to return to Tokyo.

---

[64] I later reported to my Vice-Chancellor on my feasibility study and also that I could not accept the proposal to build a Japanese garden. The cost for having a campus in Japan exceeded his expectation. The matter was dropped before it was even put up for discussion. Also I heard some rumblings from another School opposing the Japanese garden on the grounds that it would spoil the environment of the campus. I did not respond to a non-event.

[65] Brisbane elected a new Mayor; the plan to link the two cities fell through.

My last official engagement was to be a guest of honour at the first reception of the Japan Australia and New Zealand Teachers' Association and to address the gathering. The day before I left Keio University I attended a farewell lunch given by my host in my honour.

In 1976 I was elected a Fellow of the Australian Academy of Humanities. Fellows of the British Academy, who resided in Australia at that time, first formed this academic body. In the same year *Science and Civilisation in China* volume 5, part 3, in which I collaborated with Needham, was published.

The gunpowder epic was never off my mind during all this time in spite of the many handicaps I had to overcome. I felt I was working at Griffith University with one arm tied behind my back. I delegated to my deputy more responsibilities than the other three School Chairmen and had to bear the consequences personally. I kept up my spirit by telling myself that in the academic world a book would be remembered much longer than years of administrative work. I made the resolution to complete the gunpowder epic before the expiry of my term of office as School Chairman. My sojourn in Japan had provided me a good opportunity to write. I went to Canberra during university vacation to use the library facilities at the Australian National University, but was recalled back to act for the Vice-Chancellor, who was admitted to hospital for a heart condition. As the Vice-Chancellor was away, his secretary also took leave. I had to take along my School secretary to work in the Vice-Chancellor's office. On the Vice-Chancellor's desk I completed writing the transmission section of the gunpowder epic. On my Vice-Chancellor's return to office I told him that in his absence, I had written a letter on his behalf to congratulate the Registrar on the birth of a son and asked my secretary to remove all the ashtrays from his office, besides writing the most important section of the gunpowder epic. He did not say a thing. Being a person with a strong will, he had no difficulty in giving up cigarette smoking. He had gone to hospital during a quiet season, when all the students and many members of staff were out of campus. I was not at fault for not having to do much administrative work. His office, on the other hand, was dedicated only to administration and not research. He did not know how to respond to my using his office to write the gunpowder epic.

As for my publications, I wrote only one in the Chinese language for Singapore and all the rest were in English.[66] In 1977 Canberra published a lecture I gave at the Australian National University on the history of Chinese astronomy in the form of a book.[67] On 27 February 1977 I presented a paper at a conference on cultural conflicts held at Tokyo University as a guest of the Japanese Ministry of Foreign Affairs.[68] I then went to Kyoto to

---

[66] See Ho, P.Y. (1973), "Doctors take a New Look at Acupuncture", *Hemisphere*, 17:10–15; Ho, P.Y., Lim, B. and Morsingh, F. (1973), "Elixir Plants", in Sivin, Nathan and Nakayama, Shigeru, *Chinese Science* (Camb. Mass.),153–202; Ho, P. Y. (1973), "The Search for Perpetual Youth in China, with Special Reference to Chinese Alchemy", *Papers on Far Eastern History*, 7:1–20; Ho, P.Y. (1973), "Magic Squares in East and West", *Papers on Far Eastern History*, 8:115–141; Ho, P.Y. (1974), "Chinese Scientific Terminology", *Papers on Far Eastern History*, 9:1–14; Ho, P.Y. (1974), "Higher Education in China", *Proceedings of the University of Adelaide Centenary Celebrations: Seminar on Asian-Australian Educational and Cultural Relations, Past, Present and Future*, 101–109; Ho, P.Y. (1974), "Chinese Alchemical and Medical Prescriptions — a Preliminary Study", *14th International Congress of the History of Science Proceedings No. 3* (1974), 295–298; Ho, P.Y. (1976), "Three Dialogues in Science", *Hemisphere*, 20:14–21, reprinted as "An East-West Dialogue on Science", *Asia Magazine* (12 December 1976), 13–21; Ho, P.Y. (1976), "Huiyi Lien Shih-sheng xiansheng 回忆连士升先生 ", in Lien Wenssu and Lien Liang-ssu, *Festschrift in Memory of Lien Shih-sheng*, (Singapore), 40–51; Ho, P.Y. and Wang, L. (1976), "On the *Karyû kyô*, the Fire-Dragon Manual", *Papers on Far Eastern History*, 16:147–159; Ho, P.Y. (1977), "The T'ang Monk-Scientist I-Hsing", *Buddhist Studies*, 17:2–4, 12–13; Ho, P.Y. (1977), "Able and Adventurous: Navigation in the Chinese Tradition", *Hemisphere*, 21:2–9; Ho, P.Y. (1977), "The Star Move Still", *Hemisphere*, 21:22–29; Ho, P.Y. (1977), "Intra-Asian Influence of Science and Technology", *Proceedings of Seventh IAHA Conference* (Bangkok), 532–549; Ho, P.Y. (1978), "Modern Scientific Development in China", *Eastern Horizon*, 17:5–9; Ho, P.Y. (1978), "Ancient Chinese Medicine", *Hemisphere* 7:36–41.
[67] See Ho, P.Y. (1977), *Modern Scholarship on the History of Chinese Astronomy*, Australian National University *Asian Studies Occasional Paper* No. 16 (Canberra).
[68] See Ho, P.Y. (1978), "Pride and Prejudice: Science in Cultural Conflicts between Europe and China", *Proceedings of Asian Colloquim on Cultural Conflicts*, (Tokyo), pp. 1–13.

congratulate Kiyosi Yabuuti for the honour of being nominated to deliver a lecture on Chinese astronomy and the calendar in the presence of Emperor Hirohito at the Imperial Palace. Yabuuti invited me to lunch with a collaborator of Needham who was then responsible for writing the textile section, hoping that both of us would know about his progress. This collaborator had much experience in the textile industry, but was quite elderly and spoke no English. He said he was glad to be able to converse with me in Japanese and told us that owing to many outstanding commitments he hoped to start writing the following year. What he wrote would have to be rendered into English. When I reported our meeting to Needham, Peter Burbidge wrote to say that he had already found a replacement for this gentleman, who, in any case, passed away of old age shortly afterwards. In August the same year I read a paper at the 7th International Association of Historians of Asia conference in Bangkok.[69] I stopped over in Singapore on my way back to Brisbane and gave a talk at the Regional English Language Centre on "Traditional Chinese Scientific and Technological Terminology" at a staff seminar. In December 1977 I participated in the 5th Leverhulme Conference held in the University of Hong Kong.[70]

My term of office as Chairman of School was due to expire in April 1978. About a month before the due date I managed to complete the draft and send it off to Needham. Wang Ling seemed to have accepted the situation. He asked Wang Gungwu to bring me several boxes of the files he made, but these came after I had already sent my draft to Needham. Accordingly I returned those boxes to Wang Ling intact.[71]

My daughter Sook Ying obtained her first degree in science at the University of Queensland and carried on to work for a Ph.D. degree in

---

[69] See Ho, P.Y. (1977), "Intra-Asian Influence of Science and Technology", *Proceedings of Seventh IAHA Conference* (Bangkok), pp. 532–549.

[70] See Ho, P.Y. (1981), "Modern Scientific Development in China", *Proceedings of Fifth Leverhulme Conference, University of Hong Kong, December 1977*, pp. 265–286.

[71] Wang Ling later sent these boxes to the East Asian History of Science Library, Needham Research Institute, Cambridge.

chemistry.[72] I completed my term of office as Chairman of the School of Modern Asian Studies in April 1978. According to the regulations of Griffith University the term of office of a School Chairman was three years which could be renewed for another term of three years. The term of the four first School Chairmen was five years, renewable for another term of three years. The Vice-Chancellor would appoint a Chairman of School in consultation with members of the School concerned and with the approval of the university council. It so happened that all the four Schools turned out to have a new Chairman each. The university Registrar asked me to speak on the behalf of my colleagues at a university function to thank the outgoing Chairmen for their past services. As could be expected, politics within the Schools accompanied the appointment exercise, for example, the School of Modern Asian Studies, which had three professors, including me. The Vice-Chancellor would prefer to appoint a professor to be Chairman. He knew my preference for scholarly research and would prefer to have someone with more interest in administration; I had also indicated to my colleagues in the School that I would not seek another term as School Chairman. There were only two contenders for the post, but there was much politics being played. I knew whom the Vice-Chancellor already had in mind, and tried to keep myself out of what was going on. In any case the majority agreed with the Vice-Chancellor's intention. Unfortunately, my successor did not last long. After only a few months in office he died; the Chairmanship went to the other professor.

Second Visit to China

Needham wrote to ask me to go to Cambridge to go through my draft of the gunpowder epic with him. However, duty called; the Vice-Chancellor of Griffith University asked me to lead a team of 24 university members on a study tour to China as part of my duty, but entirely at my own expense. See

---

[72] She is now Principal Research Scientist with the Australian government and adjunct professor of aerospace dynamics at the University of Adelaide.

Figure 14. The Vice-Chancellor had personally led the first student study tour to China the year before. Having obtained permission for Lucy and me to visit China from the Minister of Home Affairs, Tan Sri Mohammad Ghazali bin Shafie, we embarked on our tour with 24 members in all. When we stopped over in Hong Kong, Rayson Huang asked the Dean of Arts to look after the team by fetching us from our hotel to visit the University of Hong Kong and then to one of the floating restaurants in Aberdeen for lunch, accompanied by several student representatives, who gave their Australian counterparts a practical lesson on the use of chopsticks. We visited Guangzhou, Beijing, Datong, Shanghai, Suzhou, Hangzhou and Guilin. In Beijing I asked my colleague Dr Larry Crissman to deputise for me in accompanying the team to visit the Great Wall of China and the Ming Tombs, so that I could be free to meet the scholars doing research on the

GRIFFITH UNIVERSITY

Nathan, Brisbane, Queensland, 4111. Telephone (07) 275 7111. Telegrams Unigriff Brisbane

Vice-Chancellor:
Professor F. J. Willett

Ref: G/AA 15

23 November 1979

Professor Ho Peng Yoke
8 Holdway Street
KENMORE   QLD   4069

Dear Professor Ho

I confirm that the University requested you, as a senior member of the School of Modern Asian Studies, to take the responsibility for heading a group of students and staff of the University on a tour through the People's Republic of China over the period 24 November to 20 December 1978.

The University believes it is in the best interests of both the University and the School if such tours are led by a senior and experienced scholar and I am grateful to you for accepting this duty. As you know, for reasons of University policy, I am unable to reimburse your expenses for this trip.

Yours sincerely

F.J. Willett
Vice-Chancellor

Figure 14 An "order" from my Vice-Chancellor

history of Chinese science. There I met Dr Xi Zezong 席泽宗, Director of the Institute for the History of Natural Sciences, for the first time. We had known each other through our publications on historical Chinese astronomical records almost 20 years earlier. At his request I gave an impromptu talk on research activities on the history of science outside China to members of his institute. That was my first experience in making a speech in *putonghua* in China. After the talk was over a member of the audience approached me saying that he could follow everything I said and detected from my accent that I was a southerner. I said he was perfectly right and asked him whether he could identify which part in the south I came from. As he was thinking aloud regarding the place I could have come from, starting from Hangzhou and Nanjing to the provinces of Fujian and Guangdong, I interrupted to say that it was further south, until he finally concluded that I came from Hainan Island. When I asked him to think of some more places further south, he said that it was not possible; although there were islands in the South China Sea, they were not inhabitable. On telling him that I came from further south beyond those islands, he said that I must have come from Singapore. When I informed him that I came from Australia, he said, "Comrade, you are pulling my leg".

On the more serious side, I had a discussion on the Chinese record of the 1054 supernova with Liu Jinyi 刘金沂, a colleague of Xi Zezong. I had earlier published an article in *Vistas in Astronomy*, which stated that the record in the Astronomical Chapters of the Official History of the Song Dynasty did not give the correct direction of the supernova. Liu Jinyi told me that he had found evidence that the direction given in the above record was a misprint. Xi Zezong also arranged for a group of senior researchers together with Dr Xia Nai 夏鼐 to have a meeting with me in the same afternoon at the Beijing Hotel. There we exchanged views on the state of research on the history of Chinese science and the ways to promote interest and research in the subject. I mentioned that when Needham first went to wartime China he had the ambition of rendering a helping hand to Chinese scientists. His effort in this direction had endeared him to the Chinese people long before the publication of the first volume of *Science and Civilisation in China*. After his return to Cambridge to begin work on the *Science and*

Civilisation in China project he did not forget the need to help Chinese scholars, although the field had narrowed down to the history of Chinese science. During the 1960s I witnessed on more than one occasion his openly expressing his disappointment in not seeing participants from China at international conferences. Unfortunately his commitments in Cambridge and his involvement with the *Science and Civilisation in China* project gave him little time to divert himself to such aspirations. I suggested to the meeting that we should try to fill the vacuum, not only in helping Needham to fulfil his aspiration, but also, more importantly, to do a service to Chinese scholars of the history of science. The meeting unanimously agreed on the desirability of international symposiums on the history of Chinese science. Xi Zezong and Xia Nai hoped that I would start the ball rolling.

Sabbatical Leave in Tokyo and Hong Kong

By April 1979 I had earned six months' sabbatical, subject to the approval of Griffith University. Needham urged me to go to Cambridge to sit down together with him to go through the gunpowder epic. From my past experience that would mean spending six months in Cambridge, while I estimated that it would only take less than a month for Needham to work together with me in going over my draft on the gunpowder epic, with the rest of the time spent on waiting for Needham to be free to attend to me. I had been living in a western environment for six long years so Cambridge would not give me much of a change. Instead of using my hard-earned sabbatical to go to Cambridge to join Needham I chose to take my sabbatical partly in Japan and partly in Hong Kong to have a change of environment. My decision turned out to be to the advantage of both Needham and me. As events turned out, Needham went to the Chinese University of Hong Kong for the Chien Mu lecture series, while I happened to be in Hong Kong at the same time. It was that visit of his that led to the formation of the Hong Kong East Asian History of Science Foundation. According to the plan for *Science and Civilisation in China* at that time, the gunpowder epic section was to combine with the paper and printing section to form volume 5, part 1. Even if Needham and I had set out to produce the final version of the gunpowder epic together in Cambridge in 1979, there would still be a problem when the paper and printing section by Tsien Ts'uu-hsiun 钱存训 turned out to be so substantial a few years later that it had to form one independent book. It became volume 5, part 1. Not wishing to see a thin book for the gunpowder epic, Needham enlarged what I had written to make it volume 5, part 7 all by himself.

I spent half of my sabbatical of three months in Tokyo as a Visiting Researcher at Tokyo University, attached to the International Relations Department headed by Shinkichi Eto 卫藤审吉 at the Komaba campus. I went daily to the National Diet Library. There I came across a rare handwritten copy dated 1805 from an original 16th-century block-printed edition of the alchemical work, *Danfang jianyuan* 丹方鉴原. The original 16th-century work is in the library collection of the Imperial Palace of Japan. I copied the whole book by hand, as making a photocopy of this rare book was not permitted. At my request the library made photocopies of three selected pages of the book for the purpose of illustration. While copying the book, I compared it with the modern version that was reproduced from the 15th-century edition and found it to be more complete and accurate than the modem printed version. I also made the acquaintance of the Chief Librarian of the National Diet Library, Mr Minoru Kishida 岸田实.

The other three months of my sabbatical were spent at the University of Hong Kong, as an Honorary Professor, attached to the Department of Chinese and the Centre of Asian Studies. Rayson Huang offered me free accommodation with free breakfast at the old Master Flat of Robert Black College. In return I offered to give a series of four public lectures on the history of science at the Science Faculty of his university. My lectures were to appear later as a *Griffith Occasional Paper*, published jointly with the Centre of Asian Studies.[73] I also conducted three seminars in the M.A. by coursework programme of the Department of Chinese, gave a public lecture on the influence of the *Book of Changes* on traditional Chinese science that was

---

[73] See Ho, P.Y. (1982) *The Swinging Pendulum: Science in East and West with special reference to China* (Centre of Asian Studies, University of Hong Kong).

jointly organised by the Department of Chinese and the Centre of Asian Studies on October 12, and conducted a staff seminar at the Department of Physics. At Robert Black College I worked on the material collected in Tokyo and wrote a book that I dedicated to my friend Rayson Huang.[74] My son Yik Hong had an elective term during his clinical years that required him to gain outside experience by being attached to another university or hospital during vacation time. He found attachment in the medical school of the University of Hong Kong and stayed with me in my flat at the Robert Black College.

Needham was in Hong Kong giving the series of Chien Mu Lectures at the Chinese University of Hong Kong. Lu Gwei-Djen was with him, and I had several occasions to meet them. I went over to the Chinese University of Hong Kong in Shatin to attend two of his lectures, but could not fit in the others because they clashed with my own lecture time. It was for the same reason that I could not join Needham in a television interview; Dr Thomas Lee 李弘祺 of the Chinese University of Hong Kong played the part of an interviewer instead. Needham's visit received much publicity in the local Chinese newspapers. He made known to his friends his desire to look for funding for a building in which to house his library. He had an old friend in Peter Lisowski, the Professor of Anatomy at the University of Hong Kong, at that time. Needham was the houseguest of Peter Lisowski and Ei Yoke (Mrs Lisowski) on several occasions. During his days in China Needham also knew Mr Li Tsung-ying 李宗瀛, a correspondent with the *Dagongbao* (*Ta Kung Pao* 大公报) newspaper, who became the editor of the *Eastern Horizon*.[75] Then there was Dr Peter Lee 李励生, who spent some time

---

[74] See Ho, P.Y. (1980), *Daozang Danfang jianyuan* 道藏丹方鉴原 (Tuku Tao's *Tan-fang chien-yuan*: a 10th-Century Alchemy Source-Book, in Chinese, with an Introduction in English) (Centre of Asian Studies, University of Hong Kong).

[75] Lisowski adopted a Chinese name Li Shouji 李守基 and Needham took the Chinese name Li Yuese 李约瑟. They had the same Chinese surname as Li Tsung-ying and Peter Lee. Thus the three members of the Li clan in Hong Kong put their heads together in order to help their fellow clansman in Cambridge. So went the story.

between 1976 and 1977 in Cambridge to work with Needham on *Science and Civilisation in China* while under the employment of Coco Cola. The three of them met together and decided to approach Dr Philip Mao 毛文奇 to form and be the Chairman of a trust set up for the purpose of helping Needham. Philip Mao was a well-known surgeon and art collector with a wide circle of friends in Hong Kong. This was the beginning of the East Asian History of Science Foundation, Hong Kong. At its first function, it invited prominent businessmen in Hong Kong to a reception in the presence of Needham and Lu Gwei-Djen. I was also present at the reception.

Needham made no mention of the gunpowder epic when we met each other in Hong Kong. On his return to Cambridge, however, he wrote to ask me to go to Cambridge in early 1980 to work with him on my draft. I replied politely that I would be unable to make the trip. Needham decided to work on my draft by himself.

I got back to routine teaching at Griffith University after my return to Brisbane. 1980 also saw the graduation of my son Yik Hong with a M.B., B. S. (Hons) degree from the University of Queensland. Until then Lucy and I were able to have all our five children living together with us under the same roof. I began to realise that we would not be able to maintain the same state very much longer.

Griffith University found itself operating on a very tight budget during the year 1980. It happened that the Chair of Chinese at the University of Hong Kong was about to fall vacant upon the retirement of Professor Ma Meng 马蒙, and the university was finding it difficult to find a suitable candidate to fill the vacancy. Rayson Huang negotiated with Griffith University to appoint me as Professor of Chinese and Head of Department on a three-year secondment. The secondment would begin from April 1981. There was talk of staff retrenchment in the air. I proposed that those who were able to get a secondment to universities overseas if not to other universities in Australia might render such a step unnecessary. I noticed a marked change in the School of Modern Asian Studies. I could no longer find in the School any Japanese visiting professor supported by the Japan Foundation or a visiting scholar from Southeast Asia sponsored by the Malaysian Airlines

System. Neither was there any exchange student from the Tokyo Daito Bunka University.

I heard news that Professor Ulrich Libbrecht was planning to hold an international conference on the history of Chinese science at Leuven, Belgium. I corresponded with him on the dates of his intended conference and suggested to him that, in the unlikely event that funding became a big issue, he might consider holding the conference jointly with the Department of Chinese at the University of Hong Kong in December 1981. Eventually Ulrich Libbrecht held the conference in Leuven in 1982. In November 1980 I attended the First Australian Conference on the History of Mathematics held in Monash University, Melbourne.[76]

---

[76] See Ho, P.Y. (1981), "Ancient Chinese Mathematics", *Proceedings of the First Australian Conference of History of Mathematics* (Clayton, Victoria), pp. 91–102.

The University of Hong Kong

A University Department where East and West Met

I accepted my secondment to the University of Hong Kong as a challenge. The expectation of the holder of the Chair of Chinese there was rather unique. Academically, any university in the West would be satisfied when the appointee was a Sinologist of international repute. A university in China would try to look for an eminent Chinese scholar, proven by his or her publications in Chinese, and ideally accompanied by skills in one or more of the traditional arts of calligraphy, painting, music and poetry, although such ambitions were not always realised. The University of Hong Kong is and was a Western-style university, originally modelled after the British university system. To the administration and most of the academics, other than those within the Chinese Department, a Sinologist with an international reputation would be eminently suited to fill the Chair of Chinese. However, the expectation of the Chinese community in Hong Kong was more in line with that in China. If the choice were for the latter, a head of department without proficiency in the English language would present problems to the administration of not only the Chinese Department but also the Faculty of Arts and the university as a whole. A third, but often overlooked, problem was the preference of students of Chinese in those days to listen to lectures delivered in the Cantonese dialect. Rayson Huang, who was involved around 1963 in the appointment exercise that led me to the Chair of Chinese Studies at the University of Malaya, subsequently witnessed the successful building up of the Chinese Studies Department and my acceptance by both the university and the community in Kuala Lumpur. He knew about my publications and my connections with Needham. Perhaps he had also taken into consideration that Cantonese was my mother tongue.

Before I left for Hong Kong, my friend Liu Ts'un-yan, who was knowledgeable about conditions in Hong Kong paid me a visit and talked about my secondment to the University of Hong Kong. He said that working in Australian universities demanded that we published our works in the English language. He suggested that I should publish more in the Chinese language when I went to Hong Kong. The reason he gave was to raise the standard of Chinese publications, but I felt that he was trying to advise me on how to gain acceptance by the Chinese community in Hong Kong. From personal observation as a Visiting Professor in 1979, I knew that publishing research work in the Chinese language could be a handicap for some members of the Chinese Department in the University of Hong Kong. Many members of the promotion committee or staff committee in the university hierarchy were neither able to understand the publications in Chinese nor knowledgeable about the academic standing of the journals themselves. They would seldom give the candidate for promotion the benefit of the doubt. I resolved that, as head of department, I should do what I could to help by explaining the publications and commenting on the standing of the journals concerned.

The question of gaining public acceptance fortunately did not arise when I worked in the University of Hong Kong. In 1982 the Hong Kong Society of Translation elected me Honorary Fellow of that society. I received invitations to speak in either English or Cantonese to universities, learned societies, social clubs and high schools. I spoke several times at the United College of the Chinese University of Hong Kong, both because of the friendliness of its President Professor Chen Tien-chi 陈天机 and because Lucy's uncle, Sir Kenneth Fung, was a past Chairman of the Board of Trustees of the College.[77] I spoke on a variety of topics, from Daoism to an audience of

---

[77] For example, see Ho, P.Y. (1984), "Li Madou jiqi zai keji zhuanbo shang suo ban jiaose 利马窦及其在科技传播上所扮角色" (Matteo Ricci and the part he played in the transmission of science, in Chinese), *United Bulletin,* 40:3–5.

Roman Catholic nuns to Chinese mathematics at the St. Stephen College.[78] I cannot forget the occasion when I was the guest speaker on Speech Day at the La Salle College in Kowloon and had to give a talk on the importance of learning English. One might wonder why the Principal picked the Professor of Chinese instead of the Professor of English to speak on this topic. The reason was simply that we knew each other in the late 1940s when we were in the same school in Ipoh, Malaysia. On the morning of the Speech Day I had breakfast at Robert Black College together with a guest from North America. On hearing that I was about to speak on the importance of learning English, my guest said jokingly that it would be a very easy thing for me to do. He suggested that I just needed to say the following:

> "I am the Professor of Chinese. Hong Kong will soon be returned to China. There is no future for English. You should learn Chinese instead. Are there any questions?"

I did begin my speech at the La Salle College with these words, saying that it was the advice given to me by an American professor on his visit to Hong Kong, but I added that neither was this advice made seriously nor would I take it to let down my old friend from Ipoh, the College Principal. The American professor had unwittingly provided me the means to start off my talk with something unexpected to capture the attention of my audience.

Chinese newspapers and magazines frequently asked me to contribute articles. I wrote for the *Ta Kung Pao*, the *Sing Tao Daily* (星岛日报) and the *Ming Pao Monthly Magazine* (明报月刊), to mention a few. Interestingly, the local Chinese press carried different versions of my native home. One said that it was Panyu district, another said Shunde 顺德 district, while a third and a fourth said it was Shantou city and Fujian province. I made no response to these conflicting versions; I could not feel any difference after all. Understandably, being a Professor of Chinese, I received less attention from

---

[78] The word "college" was used in a narrower sense in Hong Kong than in Britain, generally confined to a sub-university tertiary institution or a secondary school (sometimes with primary classes).

the local English newspapers. However, the leading English newspaper was keen to carry views by a Professor of Chinese on popular issues outside his specialty. The administration office of the University of Hong Kong seemed to take an interest in the activities of its staff which were reported in the English press. Figure 15 from the *South China Morning Post* is an example of a newspaper cutting I received from that office with its compliments. Perhaps the university was glad to notice that its Professor of Chinese was not living in an ivory tower.

One important thing I realised that one had to handle well before I assumed duty in Hong Kong was staff matters. The Department of Chinese in the University of Hong Kong then probably had the largest number of students and largest number of teaching staff among the universities outside China. Teaching was divided into three groups, namely Chinese literature, Chinese history and translation, and the teaching staff naturally fell into these groups. Competition was inevitable, especially in the case where the head of department belonged to one of the groups and was suspected, whether or not justifiably, by the other two groups to show favouritism to his own. My policy was to show a personal interest in all the three groups and my respect for the individual expertise of every member of my colleagues. My own expertise in the history of science was a great asset,

A living language must reconcile to evolution, change

I READ the two letters of Colin Porter Pountney (SCM Post, January 14) and Clarence McCord (SCM Post, January 26) with interest.

Perhaps one may like to hear from a non-native speaker of English who has no axe to grind with either side of the Atlantic and no expertise in their languages either.

Sentimentally, I tend to share the view of Mr Pountney.

I hate to have to unlearn what I was taught in school several decades ago. This is rather disconcerting for one whose mother tongue is Chinese.

However, I have to accept a fact of life that evolution continuously takes place in all living languages in the world.

For example, each time I visit Tokyo I find that I have to acquire a new vocabulary.

Once I ordered a plate of fried rice at a small restaurant frequented by undergraduates of Keio University and asked for "yaki meshi" and when the attendant asked me once more, I said "cha han," both terms meaning "fried rice." But "Hai, furaido raisu" came the reply.

Some Japanese may regard this as an English or American threat to the Japanese language. However, in making judgment I must not let sentiment get the better of me.

I see the problem not as a question of threat or no threat but rather as the vitality of English in its role as the most commonly used language in the world.

This is, indeed, the greater contribution by Britain to human civilisation.

I can perceive two camps, one considering the American threat to English as real, and the other holding the opposite view. Which direction the English language will follow depends on how much each side can assert itself or hold its own.

The situation does not remain static — usage alone determines whichever individual case will eventually prevail.

This is a natural phenomenon for living languages such as English and Chinese, and their magnitude grows with the number of people using them.

Such a phenomenon becomes rare only in the case of a "dead language" like Latin.

HO PENG-YOKE
Professor of Chinese
University of Hongkong

Figure 15 The *South China Morning Post*

as it offered no threat to the expertise of any individual and was beyond challenge in that part of the world. Just as in Kuala Lumpur, it was fortunate that I did not play *mahjong* nor drink strong alcohol at dinner. I did not join my colleagues in such ventures to avoid becoming too familiar with any individual or being suspected of having been so. However, I was always approachable and I left my office door wide open so as to be transparent.

I thought of making some special contribution to the Department of Chinese during my term of office in Hong Kong. Developing a course on the history of Chinese science alone would not be of general significance to the whole field of Chinese studies. I saw the problem of communication between traditional Chinese scholars and Sinologists. This had not been a problem to me since my days in Kuala Lumpur, nor to some of my colleagues at the University of Hong Kong. My experience in Japan, however, brought in a new dimension. Japan had a very important part to play both in the preservation of Chinese texts that were already lost in China, and in Chinese studies itself, for example, writings on Chinese literature and philosophy. I noticed that most Japanese scholars in Chinese did not speak the modern *putonghua* official common Chinese dialect and read classical Chinese texts in an ancient Chinese local dialect with punctuation and clarifications added in Japanese — they read Chinese texts as *kambun* 汉文. Reading in *kambun* makes a text easier to understand to a Japanese student than reading the same text in Chinese to a Chinese student. Since many, but not all, Japanese professors of Chinese did not speak *putonghua* or write in English or Chinese, I wrote in Japanese in my official correspondence with them.[79] Figure 16 shows an introduction written for a Japanese book on Dunhuang music.

---

[79] I must add that there were also Japanese professors in Chinese who spoke impeccable *putonghua*. I had a colleague who was educated in and received a Master of Arts degree from Kyoto University. I had not heard him speak Japanese and was wondering why he did not learn Japanese during his stay in Tokyo. Of course, not having heard him speak Japanese did not necessarily mean that he could not speak Japanese. One day I met his teacher Professor Shigeru Kyomizu 清水茂. On conversing with him in *putonghua* it dawned on me that he was in Kyoto to teach my colleague modern Chinese literature and not the Japanese language.

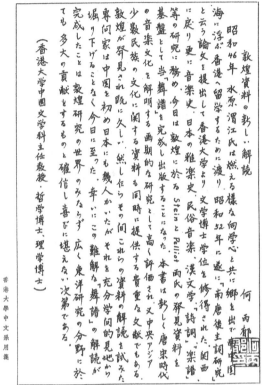

敦煌資料の新しい解読

昭和46年水原渭江氏は燃える様な向学心と共に郷を出で海に浮ぶ香港へ留学するために渡り昭和32年に逐に「南唐後主詞研究」と云う論文を提出して香港大学より文学博士学位を修得された。関西に戻り更に音楽史、日本の雅楽史、民俗音楽、漢文学、詩詞、楽譜等の研究に務め、今日は敦煌に於る Stein と Pelliot 両氏の発見資料を基盤として当「舞譜」を完成し出版することになった。本書は新しく唐来時代の音楽文化を解明する画期的な研究として高く評価され、又中央アジア少数民族の文化に関する資料も同時に提供する貴重な文献でもある。敦煌が発見され既に久しい。然し乍らその間これらの資料の解読を試みた専問家は中国を初め日本にも幾人かいたがそれを充分学問的見地から掘り下げることなく今日に至った。幸いにこの難解な舞譜」の解読が完成したことは敦煌研究の世界のみならず広く東洋研究の分野にても多大の貢献をするものと確信し喜びに堪えない次第である。

（香港大学中国文学科主任教授・哲学博士、理学博士）

何丙郁

香港大學中文系用箋

Figure 16 Introduction to a Japanese book

Since I had colleagues in the Department of Chinese who were educated in Oxford as well as Australia, I would like to see some colleagues with Japanese experience, especially when Japan was a nearer neighbour to Hong Kong than the Western countries. I encouraged and helped a student of mine doing history of Chinese science, Fung Kam Weng 冯锦荣, to study in Kyoto under Keiji Yamada 山田庆儿. He obtained a doctoral degree from Kyoto University and is now teaching the history of Chinese science course that I started at the University of Hong Kong. At the request of a colleague, Lai Wood Yan 黎活仁, I became his supervisor for his Ph.D. candidature on modern Chinese literature. He had earlier received his Master's degree from Kyoto University. The reason for my agreeing to be his supervisor was my desire to see some of my colleagues with Japanese experience. Although it was a case where the student had much more knowledge of the subject than his supervisor, Wood Yan completed his candidature in good time and submitted a highly commendable thesis.

Conference on the History of Chinese Science and Needham in East Asia

I assumed duty at the University of Hong Kong in April. After only a few days at the Robert Black College, I moved to an apartment at Fulham Garden, Pokfulam Road, about ten minutes' walk from the university. My office was in a corner on the top floor of the Loke Yew Building. Soon after my arrival in Hong Kong I received bad news from Kuala Lumpur that my younger brother Ho Thong had passed away in the University Hospital there. I last saw him in Sydney in 1980.

I gave a public lecture at the Space Museum on 29 April as one of the Hong Kong-Denmark Lectures on Science and Humanities, sponsored by the Carlsberg Research Establishments.[80] In August 1981 I went to the International Expo in Osaka, accompanied by my *Ippōkai* friends at their annual gathering. I then went to Tokyo to discuss with Shigeru Nakayama about the 1983 2nd History of Chinese Conference that I was planning to hold in Hong Kong. Nakayama took the opportunity to take me to a meeting concerning Needham's forthcoming trip to Japan to launch the publication of the Japanese translations of the four volumes of *Science and Civilisation in China*. The publisher decided to end the publication at this stage as it found the Japanese version to be a financial liability, taking the absence of volume 5, part 1 as the excuse for not being able to continue the translation. Nevertheless, the Japanese version must not end without ceremony. There were only two other persons at the meeting, which was held over lunch in a Chinese restaurant at the Okura Hotel, Tokyo. They were Notamauhiko Katayama 片山宣彦, the publisher of Shisakusha 思索社, and Teruyo Ushiyama 牛山輝代, Professor of English at a university of music who acted

---

[80] See Ho, P.Y. (1983), "Tycho Brahé (1546–1601) and China" in *Hong Kong-Denmark Lectures on Science and Humanities* (Hong Kong University Press), pp. 53–60.

as interpreter for Needham and Lu Gwei-Djen during their visits to Japan. They discussed how to look after Needham. I suggested that they provide a big car, for Needham to take the front seat and to carry his substancial luggage in the boot.

On 11 October Needham arrived in Hong Kong accompanied by Lu Gwei-Djen on their way to Japan. By then my son Yik Hong had come to Hong Kong to work as a medical officer. He used to fetch Needham and Gwei-Djen in his car. Needham and Gwei-Djen stayed with Peter Lisowski in his university quarters. It was at Lisowski's home that Needham heard the news from Cambridge that the East Asian History of Science Trust would have a research institute named after him. The next day Philip Mao organised a reception party at the Peking Restaurant, Alexander House, in Hong Kong to launch the formation of the East Asian History of Science Foundation, Hong Kong. On 14 October he held a reception at the City Hall to publicise the Foundation and to attract donors. Needham and Lu Gwei-Djen returned to Hong Kong from Japan in November. At my invitation Needham came to the Department of Chinese to give a talk on his methodology to the M. A. by coursework students. I was then at the point of completing my writing of the final draft of *Zhongguo kejishi gailun* 中国科技史概论 together with a colleague Ho Koon Piu 何冠彪.[81] We decided to dedicate the book to Needham and Lu Gwei-Djen to mark the occasion of their visit. I informed Needham about the 2nd Conference on the History of Chinese Science to be held in December 1993 in Hong Kong, and he agreed to be present.

In 1981 the Hong Kong government appointed me as a member of its Chinese Foundation Advisory Board. Following the retirement of Peter Lisowski from his Chair of Anatomy of the University of Hong Kong and his migration to Tasmania, I filled his vacancy in the East Asian History of Science Foundation, Hong Kong. 1981 was also the year that my eldest daughter completed her studies at the medical school of the University of Queensland. She later went to London to do her M.R.C.P. examination,

---

[81]See Ho, P.Y. and Ho, K.P. (1983), *Zhongguo kejishi gailun* (Outline of History of Science in China, in Chinese), Zhonghua Book Co., Hong Kong.

followed by an M.D. degree from the University of Melbourne and a Fellowship of the College of Physicians.

In the spring of 1982 I went to Cambridge to attend the "Geo-Cultural Visions of the World Seminar" organised by the United Nations University at Cambridge with the help of the East Asian History of Science Library at Brooklands Avenue, Cambridge. Needham showed me the final version of the gunpowder epic section that he had completed on his own. After he had worked on it, my original draft doubled in size. This was due to a change of plans. Since the paper and printing section by Tsien Ts'un-hsiun, originally planned to be combined with the gunpowder epic section to form part 1 of volume 5, turned out to be so substantial as to occupy the whole of part 1, the gunpowder epic section needed to be enlarged to form an independent volume as part 7 of volume 5. Lu Gwei-Djen complained to me privately that Needham had spent too much time on the gunpowder epic section. I could read that she was worried about the section on medicine that Needham was supposed to write together with her. I saw Dorothy Needham before I left Cambridge. I later received a letter from her (Figure 16). That was the last letter she wrote to me.

In April I gave two staff seminars at the Department of Chinese and the Department of Philosophy of Zhongshan University in Guangzhou. That gave me my first experience of giving a formal lecture at a university in China. I took the opportunity of my visit to establish relations between the Department of Chinese of the University of Hong Kong and its counterpart in Zhongshan University. Subsequently there were frequent visits by members of the academic staff between the two universities. There were then two students from Griffith University studying Chinese over there. I took them to dinner at the Nanyuan Restaurant.

I helped the Dean of Arts of the University of Hong Kong to organise a conference in summer at the Australian National University on library computerisation for librarians in Mainland China and Taiwan. The Australian National University was responsible for applying for funds to bring in delegates from Mainland China, while the Dean of Arts had to do so for the delegates from Taiwan. I cannot forget the occasion when he took me out to lunch with the would-be donor, who was our host. Towards the end of the

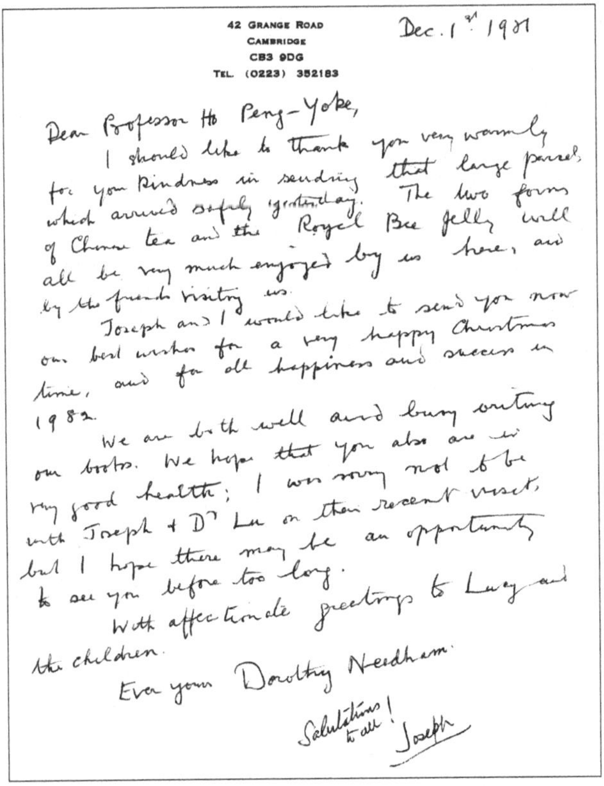

Figure 17 Letter from Dorothy Needham

lunch our host asked the Dean how much money was needed. The Dean said he needed HK$80,000. Our host took out his chequebook and said, "For such a small amount of money you need not have troubled Professor Ho by bringing him along." Perhaps that was the Hong Kong way of doing things. I went to Canberra to attend the conference and spoke to give thanks to the Vice-Chancellor of the Australian National University. This was one of the earliest occasions where scholars from Mainland China and Taiwan met.

I received an invitation to present a paper at Zagreb, Yugoslavia, in September that year. It was a conference organised by the International Association of Universities in Paris. Rayson Huang was then the President of the Association of the Southeast Asian Higher Institutes of Learning (ASAHIL). He asked me to represent him as the observer from ASAHIL to save him the time and trouble of making a long trip. Thus I attended the conference wearing two hats. The paper I delivered was about the relationship between the humanities and science.[82] There was much discussion on the bridging of the gap between humanities and science, and the conference agreed generally that the teaching of the history of science was one of the methods. One member singled out Needham as one who could bridge the humanities and science.

Spanning the humanities and science had been the main theme of my talks and writings in my early days in Hong Kong. For example, in my inaugural lecture in 1982 I illustrated this theme by the philosophy of neo-Confucianism.[83] In 1987 I wrote an article on the link between science and the humanities for the *Ming Pao Monthly Magazine.*[84] In the same year the Taiwanese magazine *Kuo-wen tien-ti* 国文天地 reproduced my article on "Science and the Humanities".[85] My research, however, did not restrict itself to this main theme. 1982 witnessed the 77th birthday of Kiyosi Yabuuti, an important occasion in Japanese life. I contributed an article in Japanese to a *Festschrift* in his honour.[86] I had earlier contributed an article to a *Festschrift* in honour of Needham's 80th birthday, but the article was only published in 1982.[87] 1982 was also the year when the Fung Ping Shan Library celebrated

---

[82] See Ho, P.Y. (1983), "The Special Place of the Humanities in Adapting Traditional Values to Changing Scientific and Technological Patterns and their Impact on Contemporary Society", in International Association of Universities, *Contemporary Scientific and Technological Changes: Their Impact on the Humanities in University Education* (Paris), pp. 21–24.

[83] See Ho, P.Y. (1982), "In Harmony with Nature: Principles spanning the Sciences and the Humanities", *University of Hong Kong Supplementary to the Gazette*, 29.4, all.

[84] See Ho, P.Y. (1987), "*Kexue wenxue yixian qian* 科学文学一线牵" (A Thread linking science and literature, in Chinese), *Mingbao yuekan* (Hong Kong, March 1987) pp. 84–91.

[85] See Ho, P.Y. (1987), "*Kexue yu wenxue* 科学与文学" (in Chinese), *Kuo-wen tien-ti*, 2.11:68–75.

[86] See Ho, P.Y. (1982), "*Chikyōzu no kenkyū* 地镜图の研究" (On the *Chikyōzu*, in Japanese), *Festschrift in honour of Professor Yabuuchi Kiyoshi* (Kyoto), pp. 143–153.

[87] See Ho, P.Y. (1982), "*Zaohua zhinan de yanjiu*" (On the *Zaohua zhinan*, in Chinese), in *Explorations in the History of Science and Technology in China*, (Shanghai) pp. 357–366.

its Golden Jubilee. The library had the largest collection of Chinese books among all the British Commonwealth universities. As Head of the Department of Chinese I had close official connections with this library, while in my private capacity, the donor of the library was Lucy's grand uncle and my father-in-law was his trusted assistant in business. I was glad to be able to contribute an article in celebration of the Golden Jubilee of the library.[88]

Professor Liu Kwang-ting 刘广定 from the Taiwan National University saw me at my office and asked me to visit Taiwan to promote the study of the history of Chinese science. His visit was followed by an official invitation from Professor Wang Shou-nan 王寿南 of the National Chengchi University. On 20 December Lucy, together with our two youngest daughters, Sook Kee and Sook Pin, joined me on a tour to Taiwan. My first assignment was to give a talk at the Institute of Modern History of the Academia Sinica. The topic of the talk concerned my connections with Needham. Two important events resulted from this occasion. During question time a member of the audience asked me why Needham had never visited Taiwan. I replied asking whether Taiwan had ever sent him an invitation. This paved the way for Needham's visit to Taiwan in 1984.[89] The *Bulletin of the Institute of Modern History* published the whole text of my talk.[90] Magazines and newspapers in Hong Kong and Singapore quoted from this article. The

---

[88] See Ho, P.Y. (1982), "*Keji wenxian jicun* 科技文献辑存" (Restoration of some scattered fragmentary Chinese texts on traditional science and technology, in Chinese), *Fung Ping Shan Library Golden Jubilee Essays*, vol. 1 (Hong Kong), pp. 124–140.

[89] Wang Shou-nan and Liu Kwang-ting knew Needham's old friend, Mr Chen Li-fu 陈立夫, who was the Minister of Education when Needham was in wartime China and who supported the translation of *Science and Civilisation in China* into Chinese in Taiwan. Chen Li-fu was still influential in the government in those days. I also wrote to Needham saying that his friends in Taiwan would welcome his visit.

[90] See Ho, P.Y. (1983), "*Wo dui Li Yuese he Zhongguo kejishi de renshi* 我与李约瑟和中国科技史的认识" (Joseph Needham and History of Science as known to me, in Chinese), *Bulletin of the Institute of Modern History, Academia Sinica* (Taipei), 1.12:425–438.

manager of the Hong Kong branch of Sanlian Book Company approached me to write a book on Needham. Lu Gwei-Djen told me that she and Needham read about my article in a Chinese newspaper while they were in Singapore. Although I had visited Taiwan several times before, my talk at Academia Sinica was my first academic contact with Taiwan. I also gave lectures at the National Taiwan University, the National Chengchi University and the National Normal University. There was little activity in history of science research in Taiwan in those days. Apart from Liu Kwang-ting the only other person with an interest in the subject that I met was Wang Ping 王萍 of the Institute of Modern History, Academia Sinica. After my official commitments my family and I went on a tour to the Sun Moon Lake and Kaohsiung, accompanied by Mrs Su, the wife of a former colleague in Kuala Lumpur, Su Ying-hui.

After her return to Australia, Sook Kee successfully completed her medical school examinations in 1983 with an M.B., B.S. degree. 1983 also saw the publication of two books in Chinese that I wrote in collaboration with my colleagues. One was the *Zhongguo kejishi gailun*, which was dedicated to Needham and Lu Gwei-Djen. A pirate edition of this book later appeared in Taiwan. The other was the *Ningwang Zhu Quan jiqi Gengxin yuce* 宁王朱权及其庚辛玉册 (Prince Zhu Quan and his *Gengxin yuce*) with L.Y.Chiu, which was published as *Griffith Asian Papers* Monograph No. 8 in Hong Kong.

In August 1983 I attended the 31st International Congress of Human Sciences in Asia and North Africa in Tokyo and Kyoto. I met Prince Mikasa, a younger brother of Emperor Hirohito, and the Nobel Laureate Hideki Yukawa 汤川秀树 at the opening reception held at the Akasaka Prince Hotel. Prince Misaka had been at the Australian National University before, and hearing that I had also been there, he shook my hand vigorously like old friends meeting each other, much to the surprise of the delegates around. I read about Yukawa in my student days. He had two brothers who were eminent scholars in Chinese. I have met all the three brothers. In Tokyo I was interviewed by NHK Television. A friend later wrote to say that I appeared in Channel 3 of the NHK *Kyōiku* Television programme entitled *Ajia no me: Sekai no me* アジア の 目 : 世界の目 on 20 September that year.

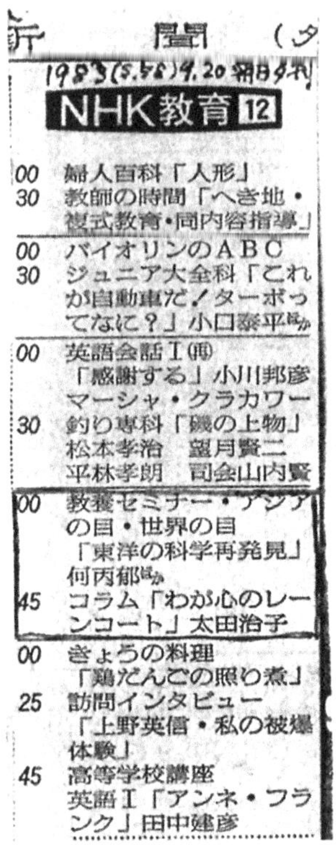
新聞（夕
1983(S.58)9.20 朝日夕刊
NHK教育 12
00 婦人百科「人形」
30 教師の時間「へき地・複式教育・同内容指導」
00 バイオリンのABC
30 ジュニア大全科「これが自動車だ！ターボってなに？」小口泰平ほか
00 英語会話I(再)「感謝する」小川邦彦 マーシャ・クラカワー
30 釣り専科「磯の上物」松本孝治 望月賢二 平林孝朗 司会山内賢
00 教養セミナー・アジアの目・世界の目「東洋の科学再発見」何丙郁ほか
45 コラム「わが心のレーンコート」太田治子
00 きょうの料理「鶏だんごの照り煮」
25 訪問インタビュー「上野英信・私の被爆体験」
45 高等学校講座 英語I「アンネ・フランク」田中建彦

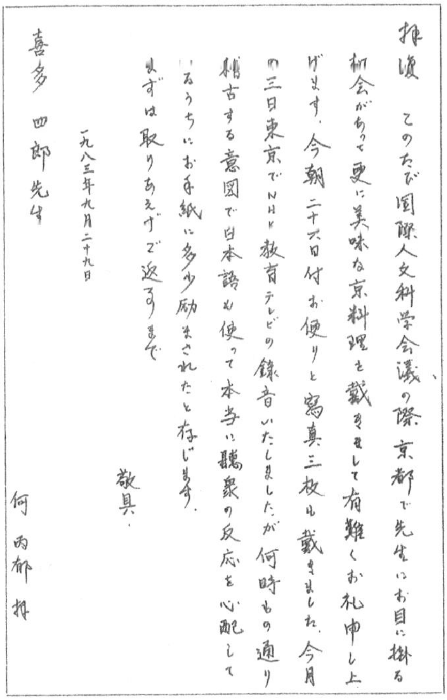
第 頁
拝復
このたび国際人文科学会議の際、京都で先生にお目に掛る
機会があって更に美味な京料理を載きまして有難くお礼申し上
げます。今朝二十六日付お便りと写真三枚も載きました。今月
の三日東京でNHK教育テレビの録音いたしましたが何時もの通り
稽古する意図で日本語を使って本当に聴衆の反応を心配して
いるうちにお手紙に多少励まされたと存じます。
まずは取りあえずご返事まで
敬具
一九八三年九月二十九日
喜多四郎先生
何丙郁 拝
香港大學中文系用箋

←Figure 18 NHK television education programme

The second part of the conference took place in Kyoto. Kiyosi Yabuuti was the organiser for the sessions on the history of science. Besides presenting my paper on the astrological interpretations of clouds and vapours in a Dunhuang MS, I shared with Keiji Yamada the responsibility of chairing the morning and afternoon sessions of section 9 of the conference on 3 September.[91]

My visit to Kyoto sparked off a new line of research for me. It happened quite accidentally. I was staying at the Kyoto Hotel. After the closing ceremony on the last day of the conference I returned to my hotel and felt that I needed a haircut. I told the front desk that I would be in the hotel

[91] See Ho, P.Y. (1984), "The Dunhuang MS *Zhan yunqi shu*", *Proceedings XXI International Congress of Humanistic Science in Asia and North Africa* (Tokyo), I:468–469.

barbershop just in case there was any telephone call for me. As my barber was about to put the finishing touch the telephone rang for me. It was a call from Michio Yano 矢野道雄, saying that he would call within half an hour to guide me to the Kyoto Royal Hotel, where Yabuuti had invited those who helped in the history of science sessions to dinner at a Chinese restaurant. Actually, the Kyoto Royal Hotel was just opposite the Kyoto Hotel. On reaching the Kyoto Royal Hotel, my attention was drawn to a small group of women fortune-tellers applying a Chinese fate-calculation method to tell the fortune of a client. That reminded me of the horoscope made for my father by a fate-calculator in Hong Kong. I decided to find out more about this method. That was how my interest in fate-calculation began. It was also a coincidence that Yano himself later turned out to be a scholar of Arabic, Indian and Chinese astrology. He was one of the last three disciples of Yabuuti. Shigeru Nakayama, an earlier student of Yabuuti, also had some interest in East Asian astrology. I had many occasions later to discuss with both Yano and Nakayama on matters concerning their expertise.

Needham went to Hong Kong in October to receive an honorary degree from the Chinese University of Hong Kong in Shatin. Shortly before I set off for Japan for the above mentioned conference, both he and Dr Ma Lin 马临, the Vice-Chancellor of the Chinese University of Hong Kong, tried to persuade me to bring forward the date of the coming 2nd International Conference on the History of Chinese Science from December to October so that Needham need not have to make another trip to Hong Kong in just under two months. I could not comply with their request because the delegates needed time to prepare their papers and make travel arrangements, and my colleagues at the University of Hong Kong were working on a tight schedule to prepare for the conference. I was an invited guest at the conferment of an honorary degree on Needham. I told Needham that I should be congratulating Ma Lin rather than him for succeeding in getting him to come to honour the Chinese University of Hong Kong by his presence.

On 20 October Needham delivered the first East Asian History of Science Foundation of Hong Kong Lecture at the University of Hong Kong. His lecture entitled "Gunpowder as the Fourth Power: East and West" was culled from the gunpowder epic section that had yet to go to the press. Rayson Huang introduced the speaker, while I said words of thanks on behalf of the East Asian History of Science Foundation, Hong Kong, as well as the University of Hong Kong. I was also responsible for editing the lecture and getting it published as a book by the University of Hong Kong Press. At a meeting with the Hong Kong Trust Needham stated that the Needham Research Institute was looking for a donation of ₤1,000,000 of which the Croucher Foundation of Hong Kong had contributed £250,000 as an endowment. Rayson Huang was a Founding Member of this Foundation. In 1983 Rayson Huang had also appointed me as a member of the University Publication Committee.

In December 1983 I organised the 2nd International Conference on the History of Chinese Science at the University of Hong Kong. Among the scholars I invited to Hong Kong were Dr Xia Nai, Professor Xi Zezong and 20 other historians of Chinese science. Although Needham cancelled the trip he had earlier promised, nobody blamed him for it. However, it was another one of the many cases of "man proposes, God disposes", with the Chinese equivalent *"moushi zai ren chengshi zai tian"* 謀事在人成事在天, in matters relating to Needham that I experienced. Originally Philip Mao intended to exploit the publicity arising from the presence of the many famous scholars from China to launch a big drive to appeal for donations on behalf of the Needham Research Institute in Cambridge. Instead of having a public function to appeal for donations, the Hong Kong East Asian History of Science Foundation held a welcoming reception for the conference delegates. At the opening ceremony Dr Francesca Bray read a congratulatory message from Needham.[92] Xia Nai in his plenary lecture on archaeology and the history of Chinese science mentioned that Needham visited the Institute of Archaeology every time he went to Beijing to find out more about recent

---

[92] Francesca contributed the agriculture sub-volume of *Science and Civilisation in China*, volume 6, part 2 (1984). That was the first book to be written independently by another author in the series. She once told me about the use of the word "collaborator" during and immediately after World War II. I think she would have preferred to be referred to as a contributor or a participant of Needham's *opus magnus*. Until very recently, she was a professor at the University of California, Santa Barbara and is now Professor of Social Anthropology at the University of Edinburgh.

archaeology discoveries. The conference operated on a very small budget with only outside donations to bring about 20 delegates from China to Hong Kong. I hosted a farewell dinner at the Alexander House Peking Restaurant for the delegates and those who helped in the conference, together with a number of my own guests, including members of the Hong Kong East Asian History of Science Foundation and Dr Cheng Te-k'un and Mrs Cheng. Subsequently, I published the papers presented at the conference in a special issue of the *Bulletin of Chinese Studies* that I established in the Department of Chinese, University of Hong Kong.

Second Period of Secondment to Hong Kong

My period of secondment to the University of Hong Kong would end by March 1984. Rayson Huang wrote to John Willett, the Vice-Chancellor of Griffith University, requesting the extension of my secondment for a further period of three years. John Willett agreed to the request. By 1984 John Willett himself had reached the age of retirement at Griffith University. Professor Roy Webb succeeded him as Vice-Chancellor of Griffith. Meanwhile Rayson Huang also appointed me Master of Robert Black College.

That year I wrote a book review for *Science and Civilisation in China*, volume 5, part 5.[93] 1984 was a memorable year for the flow of donations to the Needham Research Institute. I received a message in April from Lam Lay Yong in Singapore that Tan Sri Tan Chin Tuan had finally agreed to make a donation towards the building fund of the Needham Research Institute. (See Figure 19)

As a family friend of Tan Sri Tan Chin Tuan, Lam Lay Yong first raised the matter with him earlier in the company of his second daughter. She then arranged for Needham to stop over in Singapore to meet the donor on his way to Beijing to attend the 3rd International Symposium on the History of Chinese Science in July.[94] Accordingly Needham and Lu

---

[93] Ho, P.Y. (1984), "Joseph Needham, *Science and Civilisation in China* vol. 5, pt. 5", *Journal of Oriental Studies*, 22.1:62–63.

[94] The title of the conference had not been confirmed then.

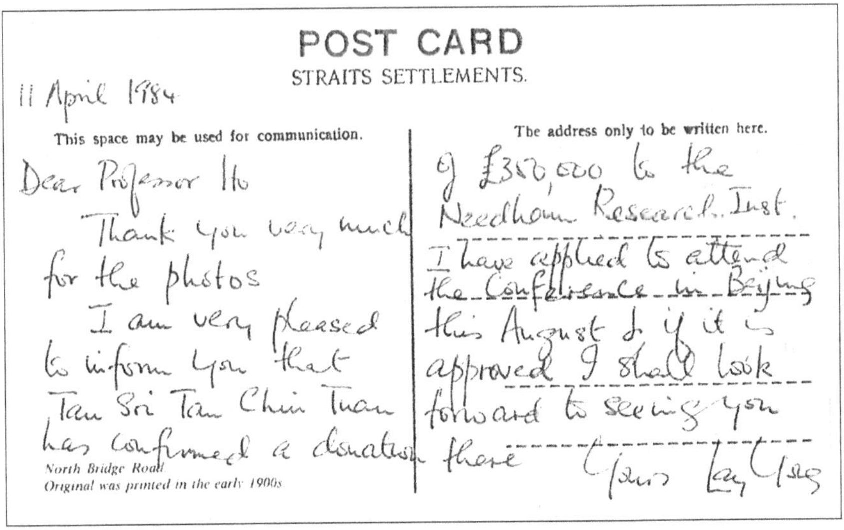

POST CARD
STRAITS SETTLEMENTS.

11 April 1984

This space may be used for communication.

The address only to be written here.

Dear Professor Ho

Thank you very much for the photos

I am very pleased to inform you that Tan Sri Tan Chin Tuan has confirmed a donation

North Bridge Road
Original was printed in the early 1900s

of £350,000 to the Needham Research Inst.

I have applied to attend the Conference in Beijing this August & if it is approved I shall look forward to seeing you there

Yours Lay Yong

Figure 19 First news about a big donation for Needham building

Gwei-Djen visited Singapore and received a donation of £350,000, the biggest individual donation to the Needham Research Institute to date. Tan Sri Tan Chin Tuan referred to a friend of his having been to Cambridge to use Needham's personal library on two occasions without having to mention his name. (See newspaper cutting from *The Straits Times* in Figure 20.) Lam Lay Yong also took Needham to see Mr Lee Seng Gee 李成义, Chairman and Director of the Lee Foundation, Singapore. Lee Seng Gee told Needham verbally that his foundation was prepared to make an annual donation of £5,000 to the Needham Research Institute.[95] That same year the East Asian History of Science Foundation in Hong Kong also sent £150,000 to Cambridge.

In July 1984 Lucy and I attended the 3rd International Symposium on the History of Chinese science held in Beijing. After the conference we went to Xi'an together with Needham, Lu Gwei-Djen and Huang Hsing-tsung. I

---

[95] This was from personal information I received from Lam Lay Yong. However, the matter seemed to have skipped Needham's mind. No contact with the Lee Foundation followed Needham's return to Cambridge until I went to Cambridge in 1990.

Mr Tan ... showed great interest in Dr Needham's project.

Dr Needham ... world renowned for his work on Chinese science.

BANKER Tan Chin Tuan will have a library at Cambridge named after him, following a donation of £350,000 (S$903,000) towards the Needham Research Institute East Asian History of Science Library building.

This £1 million (S$2.58 million) building will be the permanent home for the library's priceless collection of books and documents now housed in temporary premises on the Cambridge University campus.

Although it is not formally part of the university, the library is recognised as an associated institution.

It started with the collections, assembled since 1937, of world renowned biochemist turned science historian-Sinologist, Dr Joseph Needham, whose aim was to unravel the mystery surrounding the centuries-old history of Chinese science.

His monumental project, Science and Civilisation in China, started 40 years ago as a slim volume on the history of science, technology and medicine in Chinese culture.

It has since expanded into 20 volumes, more than half of which have already been published.

Contrary to common belief, said Dr Needham, Chinese achievements and contributions to scientific knowledge were far greater than is generally recognised.

They had science and technology long before the emergence of modern science in the West.

In Singapore last year to deliver a public lecture on Science and Technology in Traditional China, Dr Needham met Mr Tan, honorary president of the Oversea-Chinese Banking Corporation, who showed great interest in his project.

In an interview this week, Mr Tan said his donation was in recognition of the work being done by Dr Needham and his long-time chief collaborator, Dr Lu Gwei Djen, 79.

Mr Tan said a good friend had also worked in the library on two occasions and described it as a treasure house of priceless and mostly unobtainable Chinese books.

It has 10,000 books in Chinese, 40,000 in other languages, 40,000 reprints of articles and 60 journals.

"The project has been largely responsible for the increasing amount of research and publications on the history of Chinese science by scholars through the world.

"After learning about the importance of the work and the dedication of the men and women who have devoted much of their lives to it, I was naturally interested to help with the proposal to build a permanent repository for such an invaluable collection," Mr Tan said.

Informed of the proposal to name the library after him, Mr Tan had then told Dr Needham:

"...I am merely helping to provide the shell for the treasures which you and your colleagues have so painstakingly garnered through decades of research."

A plaque commemorating Mr Tan Chin Tuan's gift was unveiled yesterday in Cambridge at the stone-laying ceremony for the prestigious East Asian History of Science Library named after him.

The Chinese Ambassador to Britain, Mr Chen Zhao Yuan, unveiled the plaque which read:

"This ... commemorates the gift of many benefactors, especially that of the central building by Tan Sri Tan Chin Tuan of Singapore whose name it bears."

Dr Needham was among the 250 guests who attended the ceremony.

Figure 20 Tan Sri Tan Chin Tuan's contributions

gave a lecture entitled "The methodology of Dr Joseph Needham in writing *Science and Civilisation in China*" at the North-West University. I was introduced to the audience by Needham himself. The lecture later appeared in the *Xibei daxue xuebao* 西北大学学报 journal.[96] After my official assignment our host took us sightseeing, viewing the terracotta soldiers, the Huaqingchi 华清池 and the Beilin 碑林, the places that Needham missed out in his last visit some 40 years ago. Huang Hsing-tsung was then accompanying Needham as his personal secretary. It was interesting to watch the two of them together again. Needham was still making careful observations and taking notes like his usual self. Huang Hsing-tsung stayed close to his side wherever he went. He was then in his mid sixties; he told me he felt young again in the company of Needham, who was then approaching his mid eighties. We departed from Xi'an taking different routes to Shanghai. Needham and Lu Gwei-Djen went by way of Nanjing in order that the latter could visit her native home. Needham also made a side trip to Fujian province so that he would arrive a couple of days later in Shanghai. Lucy and I took a direct flight to Shanghai together with Huang Hsing-tsung who chose to travel with us. Thus the three of us arrived first in Shanghai, checking into at the Jinjiang Hotel. Lu Gwei-Djen arrived one day later and had a room next to ours. She was taken ill and was attended to by a doctor. Needham came the next day, and complained about the room allocated to him. It was a two-room suite on the top floor of the hotel, but Needham said that he was afraid of heights. Lucy and I volunteered to exchange rooms with him; everyone was happy. We also arranged with the hotel to bake a birthday cake for Lu Gwei-Djen. We celebrated her birthday in her hotel room. We joined Needham in meeting a number of his old friends in Shanghai. We also had dinner at the home of Cao Tianqin and his wife, Xie Xide. Xie Xide was then the President of Fudan University, Shanghai.

Lucy and I had to return to Hong Kong early to prepare for the wedding of our son Yik Hong and our future daughter-in-law Chui Wah 翠桦 (Ludmilla). Needham and Lu Gwei-Djen stayed a little longer in China

---

[96] See Ho Peng Yoke (1984), "*Li Yuese zhixue fangfa* 李约瑟治学方法*" (Joseph Needham's methodology, in Chinese), *Xibei daxue xuebao* (Xi'an), 14:1–5.

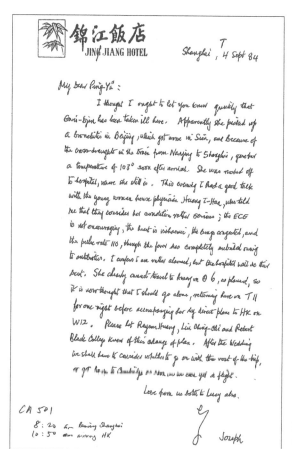

Figure 21 Letter from Needham written in Shanghai

before coming to Hong Kong to attend the wedding ceremony and the dinner party. The wedding took place on 15 September. Needham and Lu Gwei-Djen arrived a couple of days earlier and stayed at the old Master's apartment of Robert Black College as my guests. Philip Mao came to the College to see them. Needham told us that Lu Gwei-Djen had to be admitted to hospital in Shanghai after we left and feared that she might not be able to accompany him to Taiwan. Philip Mao suggested that Lu Gwei-Djen be given a thorough medical check-up and immediately made an appointment with the Professor of Medicine of the University of Hong Kong, Rosie Young 杨紫芝. Lucy volunteered to look after Lu Gwei-Djen in case she had to be left alone in Hong Kong to allow Needham to make his Taiwan trip. I took Lu Gwei-Djen to the Queen Mary Hospital to see Rosie Young, who pronounced her patient and friend fit to travel to Taiwan.

Needham and Lu Gwei-Djen apparently had a marvellous time in Taiwan. This is not the place to give an account of their travels, but only to mention that the National Palace Museum donated a set of the facsimile reproduction of the *Qinding Siku quannshu* 钦定四库全书 (Complete Library in Four Branches of Literature prepared under Imperial Edict) to the East Asian History of Science Library. Needham also received a donation of US$80,000 on behalf of his institute, which was later matched by the Chinese Academy of Science in Beijing. The foundation stone of the Needham Research Institute Main Building was laid upon Needham's return to Cambridge. The Duke of Edinburgh and the Chinese ambassador to the UK were present at the ceremony.

In December 1984 the East Asian History of Science Foundation in Hong Kong invited Nathan Sivin to give the second foundation lecture. Dr J.K. Lee 利荣康 was then the Honorary Secretary of the Foundation, replacing Peter Lee, who had been transferred to the US. J.K. Lee and I were mainly responsible for organising the lecture to be given by Nathan Sivin. I had a very happy working relationship with J.K. Lee. He was a Professor of Chemistry at the University of Oklahoma until he was recalled back to Hong Kong to look after his large family business. He was knowledgeable in both the academic and the business world. He persuaded a friend, Gordon Wu, to endow a chair at Princeton University. His work as Honorary Secretary of the East Asian History of Science Foundation was both efficient and business-like, as expected of a business executive with a personal secretary. He was in charge of the Lee Garden Hotel, which was part of his family business that dealt mainly with land and property, and founded the Hysan Foundation. I only wished that Needham and Lu Gwei-Djen had paid some attention to others besides the Chairman of the Foundation and those who hung around them. They seemed to have ignored J.K. Lee.

At a meeting of the Foundation in Hong Kong the Chairman handed to the Honorary Secretary a 1986 fund appeal brochure prepared by Cambridge that was supposed to include the names of members of the Hong

Kong Foundation, asking the latter to pass it round. Taking a glance at the list, the Honorary Secretary threw it back to the Chairman, saying that he was not involved. The Chairman then noticed that the name of the Honorary Secretary was missing from the list. He withdrew the brochure, saying that he would ask Cambridge to produce another. At another instance Needham wrote to Philip Mao asking him to book a hotel for him and Lu Gwei-Djen on the Kowloon side overlooking the harbour. Philip Mao passed the letter to J.K. Lee, who then asked his own secretary to book Sheridan Hotel for Needham, giving him the usual 10 percent trade discount between hotels.[97] J.K. Lee would have been able to give Needham a far better deal in his own hotel. It left him to ponder why Needham made that choice. Respect for Needham he certainly had, but being an academic in his own right, he could not be expected to share the layman's admiration of an academic in another field of expertise. He did only what was expected of a good Honorary Secretary. He left part of his fortune to the University of Oklahoma. It was only more than 10 years later that Sir Q.W. Lee 利国伟, a relative of the Hysan family, obtained a contribution of US$250,000 from the Hysan Foundation to support the Dragon's Ascent project of the Needham Research Institute in Cambridge and the Institute for the History of Natural Sciences in Beijing.[98]

In 1985 I published two books, one in Chinese and the other in English. I dedicated the first to Dorothy Needham and asked the publisher to transfer my royalties to the Institute for the History of Natural Sciences in Beijing.[99] I dedicated the second book, entitled *Li, Qi and Shu: An Introduction to Science and Civilisation in China,* to Needham.[100] Together with L.Y. Chiu we

---

[97] Needham and Lu Gwei-Djen stayed in the Sheridan for more than 10 days. A friend Mr George Hicks told me that when he found out from the hotel reception that their bill amounted to a five-figure sum, he quietly settled it on their behalf.

[98] See section on the Dragon's Ascent in Chapter 8.

[99] Ho, P.Y. (1985), *"Wo yu Li Yuese* 我与李约瑟" (Joseph Needham and I, in Chinese), Sanlian Book Co., Hong Kong.

[100] See Ho, P.Y. (1985), *Li, Qi and Shu: An Introduction to Science and Civilisation in China,* (Hong Kong University Press, 1985), re-published in India under another title *An Introduction of Chinese Science* (Oxford University Press in India, 1985), reprinted (2000), Dover Publications Inc.

published the first volume of *Ming shilu zhong zhi tianwen ziliao* 明实录中之天文资料 (Astronomical Records In the Ming Veritable Histories) in Chinese in Hong Kong.

The book *Li, Qi and Shu: An Introduction to Science and Civilisation in China*, however, had caused some misunderstanding with the Cambridge University Press that could be best described by the popular Japanese Buddhist term *gen ga nai* 缘がない, or *wuyuan* 无缘 in Chinese — without pre-destined relation — that invokes no blame on either party. The publisher Hong Kong University Press first sent the manuscript of this book to Peter Burbidge, leaving him to decide whether Cambridge University Press would like to publish it independently or jointly with the Hong Kong University Press. Peter Burbidge wrote to me about the manuscript, but before it was found acceptable for publication, I did not wish to inform him that the full cost of publication would be borne by a donor in Hong Kong and that I would pass on my royalties to his East Asian History of Science Trust in Cambridge. Peter Burbidge later wrote to say that he had already retired from his post at the Cambridge University Press and his successor had decided that my book would compete in sales with Colin Ronan's abridgement of Needham's *Science and Civilisation in China* that Cambridge University Press was also publishing. He then returned my manuscript to Hong Kong University Press without having referred it to a reader. I thought that the decision must have been made after consulting Needham and Colin Ronan. The Hong Kong University Press then sent the manuscript to a reader. The book received a good reception soon after it appeared. The Oxford University Press, India, sold it under the title *An Introduction to Science and Civilisation in China*. A number of universities in North America used it as a textbook. In the 1990s I heard that it was difficult to obtain in the US. All the copies had been sold out, and a Dover edition appeared in the year 2000. It was never my intention to abridge Needham's *Science and Civilisation in China*. As the title of my book showed, it was meant to draw the attention of a wider audience to Needham's work by introducing the subject itself. Cambridge and I were thinking about two opposite things. I never wasted Needham's precious time over miniscule mundane matters; I just forwarded a copy of the book to Needham since it was dedicated to him, without mentioning anything else.

The East Asian History of Science Foundation, Hong Kong, invited Dr Xia Nai to give the 3rd Foundation Lecture. In June Xia Nai wrote to me saying that he was making preparations to come to Hong Kong. However, a few days later I learned of his sudden death from his widow. The lecture was postponed to the following year, and a new speaker had to be found. The cancellation of the lecture was a great disappointment. We had hoped to take advantage of the presence of the great Chinese archaeologist from Beijing to attract the wealthy in Hong Kong to make contributions to the Needham Research Institute. It was not easy then to find suitable eminent scholars from China who would give a lecture in English. Another suitable speaker from China was Professor Ke Jun 柯俊, who delivered the 4th Foundation Lecture in 1987.

In August 1985 Lucy and I went to Berkeley to attend the 17th International Congress of History of Science. As I was then making a further study of the Dunhuang MS of the *Zhan yunqi shu* 敦煌占云气书 and attempting to reconstruct the lost fragments with the help of Ho Koon Piu, I presented a paper on this subject at the conference.[101] I gave a talk at Stanford University after the conference. We saw Lucy's eldest sister Fanny, together with her husband and members of their family in San Francisco. We were also able to meet my third cousin Thomas S. Ho 何兆明 and his family in San Jose, together with his eldest brother Samuel S. Ho 何兆孔 and his two sisters, who all lived separately in the Bay Area of California. We returned to Hong Kong by way of Japan, visiting friends in Fukuoka and Okayama. In Okayama my friend Saburo Kodera took us to see the new bridge spanning the Japanese mainland and Shikoku Island across the Seinaikai. The bridge was partly constructed by his company. At Osaka another friend took us to the sacred Buddhist Koyasan to see the temples and Koyasan University.

---

[101] See Ho, P.Y. (1985), "A long lost astrological work: the Dunhuang Ms of the *Zhan yunqi shu*", *Journal of Asian Studies*, 19.1:1–7.

In 1986 the I-wen Press in Taipei published a monograph I wrote together with Ho Koon Piu entitled *Dunhuang canjuan zhan yunqi shu* 敦煌 残卷占云气书 (The fragmentary Dunhuang MS of the *Zhan yunqi shu*). We dedicated this book to Professor Wang Shu-min for his work in textual collation and Professor Su Ying-hui for his contributions to Dunhuang studies. Together with L.Y. Chiu 赵令扬 we published the second volume of *Ming shilu zhong zhi tianwen ziliao* (Astronomical Records in the Ming Veritable Histories) in Chinese in Hong Kong. Early in the year I went to Taipei to give a plenary lecture at a conference on the history of Chinese science.[102]

Rayson Huang retired as Vice-Chancellor of the University of Hong Kong by the end of the academic year. Wang Gungwu succeeded him. In May I went to the University of Sydney for the 4th International Conference on the History of Chinese Science that I had assisted Mr Henry Chan of that university to organise. We had originally invited Dr Xia Nai to give the plenary lecture. Due to his demise, I had to give a talk on him and his work relating to archaeology and the history of Chinese science. I spoke at every official function and also helped to look after the delegates from China. The Australian Chinese Community Association in New South Wales hosted a welcome banquet for the delegates of the conference on 17 May. I thought that it was an appropriate occasion to mention Needham and the Needham Research Institute in my speech when thanking our host. The following is the text of my talk:

> The Hon. Mr Ken Gabb and Mrs Gabb, President Garry Leong and Members of the Australian Chinese Community Association, New South Wales. I must thank Mr Leong very much for his kind words and for the hospitality of your Association by inviting us to this banquet. Perhaps I may try to point out the significance of this occasion as I see it. The long history of Chinese culture is

---

[102] See Ho, P.Y. (1987), "*Kejishi yu wenxue* 科技史与文学" (History of Science and Technology and Literature, in Chinese), *Proceedings of the 1986 Conference on the History of Science* (Taipei) pp. 12–17.

something that the Chinese are proud of. But until the last two or three decades, the Chinese thought only in terms of the humanities, mainly in the areas of literature, poetry, calligraphy, painting and philosophy, and not forgetting culinary art. However, largely through the monumental work of Joseph Needham in Cambridge, it has been established that China played an important role in history — its contributions to science and technology had enabled science in Europe to develop into the world science of today. Interest in the study of the history of science and technology has grown by leaps and bounds. As you can observe from the delegates of this conference, the study of the history of Chinese science is now an international enterprise. Moreover, the study of Chinese science also finds support among the Chinese communities in many parts of the world. Those in Hong Kong as well as those in Singapore have contributed generously towards the building fund for a library in the Needham Research Institute in Cambridge. Those in Singapore have helped this conference and previous conferences with travel grants. We are much encouraged by the support given by your association to this conference. On behalf of the delegates and organisers of this conference, I wish to express our gratitude to the hospitality of your association for inviting us to this sumptuous dinner, and hope that the study of the history of Chinese science would continue to receive your support. [103]

My youngest daughter Sook Pin completed her studies at the medical school of the University of Queensland in 1986. After my return to Hong Kong I helped the East Asian History of Science Foundation, Hong Kong, to look after Shigeru Nakayama, who came to deliver a lecture on Chinese astrology as the 3rd Foundation Lecture. J.K. Lee arranged for Nakayama to stay in his Lee Garden Hotel.

---

[103] Unfortunately, there did not seem to be any follow-up from here.

In November 1986 I visited the North-West University in Xi'an and delivered two lectures at the university, besides participating in a symposium of Chinese mathematics held at that time.[104] The North-West University appointed me Honorary Professor. Unlike honorary professors in most universities outside China that refer to a temporary title given to a visiting scholar who renders non-stipendiary service at a university, an honorary professor in China is a non-stipendiary lifetime title. My host was Professor Li Jimin 李继闵, a historian of Chinese mathematics specialising in the early Chinese mathematical text *Jiuzhang suanshu* 九章算术. I had to rush back to Hong Kong to play host to Needham and Lu Gwei-Djen, who stopped over in Hong Kong to raise funds for the South Wing of the Needham Research Institute in Cambridge on their way to their last visit to China.

The South Wing of the Needham Research Institute

Back in Hong Kong I arranged for Needham and Lu Gwei-Djen to stay at the old Master's apartment of Robert Black College as my guests. I then organised a college banquet and invited some 50 guests to attend. The guests included all the members of the East Asian History of Science Foundation, Hong Kong, and their spouses, as well as some prominent members of town and gown. A talk by Needham, with an introduction by Philip Mao, followed the main meal. When everyone was having coffee, Needham stood up with the help of a walking stick and made an appeal for funds to build the South Wing of the Needham Research Institute. Immediately after the talk, Philip Mao called the members of his Foundation together and asked their views. Mrs Mary Lam said, "Those in Cambridge are so cruel that they are sending such an old man to go round asking for money. Give Dr Needham what he asks!" As there were no further views the

---

[104] See Ho, P.Y. (1987), "*Sanshiwunian de kejishiyanjiu shengya* 三十五年的科技史研究生涯" (Thirty-Five Years in History of Science, in Chinese), *Xibei daxue xuebao*, (Xi'an), 17.1:1–10 and Ho, P.Y. (1987), "*Kejishi yu wenxue* 科技史与文学" (History of Science and Technology and Chinese Literature, in Chinese), *Xibei daxue xuebao* (Xi'an), 17.2:1–9.

meeting was over within a few minutes and Philip Mao informed Needham of our decision, which was to give him £140,000 that was supposed to be matched by a similar donation from the Kresge Foundation in the US. Needham immediately called me aside and asked me to succeed him as Director of the Needham Research Institute when he retired, while making it clear that the Institute would not be able to pay me any salary. It was very difficult to turn him down at that stage. I did not really need to earn a salary at the time of my retirement. My official retirement from Griffith University was five years ahead, in the year 1991. From the look of it, Needham was still active and would give me a few more years to work in Australia before needing me to take over from him. When I agreed to his request he immediately informed Philip Mao. All the other members of the Foundation joined him in congratulating me. None of us knew then what was in store for me.

The last conference I attended during my period of secondment at the University of Hong Kong was that on "Europe and China", held at the Chinese University of Hong Kong, as Honorary Chairman of the Organising Committee. I played an active role by presenting a paper and chairing some sessions.[105] The organiser of the conference was Thomas Lee, a former student of Arthur Wright who participated in a television interview with Needham in 1979.

When I was in Hong Kong I maintained frequent contact with Cheng Te-k'un, sometimes visiting him at the Chinese University of Hong Kong or at his university quarters. He had suffered a stroke and was confined to his home in 1986. By then he had already moved out from the university quarters back to his own home in the New Territories. Thomas Lee accompanied me to visit him at his home on several occasions. One day Thomas Lee asked for my curriculum vitae saying that it was a request from a friend in the US, but revealing neither the name of his friend nor his purpose. The assurance he gave me was that his friend had a good intention, adding that

---

[105] See Ho, P.Y. (1991), "Scientific and Technological Exchanges from the sixteenth to eighteenth centuries", in Lee, Thomas H.C., ed., *China and Europe*, the Chinese University of Hong Kong (Hong Kong), pp. 189–201.

he hoped something satisfactory would result. I guessed that the friend he referred to was K.C. Chang, to whom I already owed a favour. Back in the early 1970s I asked K.C. Chang to accept one of my Malay students, Zuraina Majeed, to do a Ph.D. degree in Yale under his supervision. He willingly obliged me. Zuraina later became a professor in the *Universiti Sains Malaysia* in Penang and made her name by her discovery of the Perak Man at Lenggong. I handed the particulars to Thomas Lee as requested, but only realised two years later what the whole thing was about, when I was elected a member of Academic Sinica.

The Vice-Chancellor Wang Gungwu asked me to stay longer in the University of Hong Kong. I replied that I needed some time to wind up matters at Griffith University, and that I did not know when I would be called to go to Cambridge. My secondment to the University of Hong Kong ended on 31 March 1987.

6

Return to Griffith University

I returned to Brisbane and reported for duty the next day. I found that many things had changed. The School of Modern Asian Studies had lost its status as a division and came under the Dean of a division to which it belonged. The *Griffith Occasional Paper Monograph* series, started in 1978, was no longer in existence. Such events gave me little surprise in a university where "to be different" and "to change" were often on people's lips. I occupied myself with routine teaching, sometimes thinking ahead of what I would do in Cambridge when the time came.

While I was conducting a course on "Science, Culture and Society" at Griffith University, I heard from a student that the University of Queensland Library had acquired a copy of *Science and Civilisation in China,* volume 5, part 7 on the gunpowder epic.[106] I did not receive a copy from Cambridge, while the Griffith University Library informed me that it had stopped buying books in that series. I had to borrow the copy from the University of Queensland through a friend. I did not try to find out the reason why the Griffith University Library stopped ordering Needham's book when I noted from the title page that "University of Hong Kong" instead of "Griffith University" came after my name. I feared that I might be put to give an awkward explanation. I also gently made it known to Cambridge that I had not received a copy of the book with my collaboration, but received no

---

[106] See Needham, J. (with the collaboration of Ho Peng Yoke and Wang Ling), *Science and Civilisation in China*, vol. 5, pt. 7, Cambridge University Press.

response. I took this as another instance of *wuyuan* between the Cambridge University Press and me. Without even attempting to ask for an author's discount to buy a copy from Cambridge, I eventually obtained a copy of its Taiwan edition at only 10 percent of the price I would have to pay otherwise.

I attended the meeting of the East Asian History of Science Trust held in Cambridge on 9 September 1987. The Trust approved my appointment as Director-Designate. Needham seemed to be still quite robust then. I responded to the offer by saying that I wished Needham would allow me to enjoy carrying the title of Director-Designate for many more years to come. Needham took Lu Gwei-Djen and me to the University Arms in his car and ordered roast beef for dinner. He had lost much of his agility in parking; that was the last occasion I sat in his car.

I left Brisbane on 27 October to attend a conference on the history of Chinese science held at Kyoto to honour the 81st birthday of Kiyosi Yabuuti. This conference was by invitation only, being restricted to only 10 participants from outside Japan. Seven of them were from China; the other three were Nathan Sivin from the US, Paul S. Unschuld from Germany and me from Australia. At the opening ceremony I read a congratulatory message from Needham to Yabuuti. I presented a paper in Japanese on "Research in the history of science in China: its present and its future".[107]

After the conference I attended a reception given by *Ippōkai* members at the Heian Kaikan to congratulate me on my appointment at Cambridge. I also joined my hosts on a sightseeing tour to some of the temples in the outskirts of Kyoto. Age had taken its toll on most of the *Ippōkai* members; that was the last occasion of their assembly. I then went to Tokyo to consult some friends on word-processing software that could work in both English and Japanese, so that I could recommend a suitable one for the School of Modern Asian Studies. Shigeru Nakayama accompanied me to explore the

---

[107] See Ho, P.Y. (1992), "*Saikin Chūgoku ni okeru kagakushi kenkyū no dōkō to denbō* 最近中国における科学史研究の动向と展望" (Research in the history of science in China, its present and its future, in Japanese), in Yamada Keiji and Tanaka Tan, ed., *Chūgoku kagajushi Kaigi: Kyoto Shimpojyum Hōkokusho*, Institute of Humanistic Studies (Kyoto), pp. 164–174.

software market in Akihabara. At that time I could find only two popular systems that were capable, namely the Twin Star and the Ichitarō. My investigation suggested that in view of the cost and the problems arising from the two programs, neither should be recommended to Griffith University for acquisition.

I returned to Brisbane on 13 December. On my return I heard about the completion of the main building of the Needham Research Institute. I left Brisbane shortly afterwards for Hong Kong to attend an international symposium on Chinese studies held at the University of Hong Kong between 16 and 19 December. That was the month Dorothy Needham passed away. Needham's message bearing the sad news reached me in January the following year.

The book on Chinese fate-calculation that I wrote in Hong Kong was published in 1988.[108] I dedicated this book to Chien Mu to indicate that its approach originating from the neo-Confucian standpoint showed the logic behind the system of fate-calculation. The book was written in Chinese, but I had an opportunity to introduce the system in English at a plenary lecture at San Diego later in the same year.

Member of Academia Sinica

In July 1988 I was surprised to receive a telephone call from President Wu Ta-you 吳大猷 of the Academia Sinica at Nankang, Taipei, congratulating me on being elected a member of his academy. That was the most prestigious academic membership awarded by Taiwan, similar to the Fellowships awarded by the Royal Society or the British Academy in Britain and the Fellowships awarded by one of the Australian Academies of Science, the Humanities and the Social Sciences. Internationally it may not carry the same prestige as the Royal Society for example, but it has a very special place

---

[108] See Ho, P.Y. (1988), *Cong li, qi, shu guandian lun Ziping tuimingfa* 从理气数观点轮子平推命法 (Looking at the Ziping method of fate-calculation from the standpoint of the principles of *li*, *qi* and *shu*, in Chinese), Hong Kong University Press, Hong Kong.

among the Chinese in Taiwan. At the early stage of my career in Singapore during the 1950s, I hardly met with a Fellow of the Royal Society, except for Sir Alexander Oppenheim who was a Fellow of that Society in Edinburgh. To my contemporaries and me, a Fellow of the Royal Society would be more than qualified to be a professor in any university in our part of the world. In 1958 when my wife and I invited Joseph and Dorothy Needham, Dr David and Mrs Kate Shoenberg, and Lu Gwei-Dien to our home for dinner, I deemed it a singular honour to have three Fellows of the Royal Society among our guests. After my election to a Fellowship of the Australian Academy of the Humanities I realised that the award raised my personal academic standing among my peers in Australia and added some prestige to the university at which I worked. The Australian Academy of Humanities was an offshoot of the British Academy, being founded by Fellows of the latter Academy residing in Australia. Similarly, a Fellowship of the Royal Society, for example, would raise one's academic standing internationally and add prestige to the university where one worked. In the 1990s I talked about this issue with Sir Brian Pippard. He said that there were so many Fellows of the Royal Society and the British Academy in Cambridge that I could hardly avoid meeting one or more of them, and that a Fellow was nothing special.

By nothing special Sir Brian was referring to Cambridge, as outside Oxbridge and beyond the shores of Britain a Fellowship of the Royal Society or the British Academy still has a high international academic standing. By nothing special Sir Brian could also have been referring to the non-academic world, and what he said would apply to almost all learned societies in the world. Here the Academia Sinica is an exception perhaps as a result of the cultural differences between East and West. A member of the Academia Sinica enjoys much more prestige, and even privileges, in Taiwan than a Fellow of the Royal Society does in Britain. It did not happen to me as I did not reside in Taiwan, but if someone there was elected a member of Academia Sinica, his whole family and clan, and also the university or unit where he worked would join in to celebrate. I have seen banners and lion dances, and heard the firing of firecrackers and the beating of drums at such occasions. It brought to mind the celebration of a successful candidate in the imperial civil examinations in the bygone days of traditional China. At least at Griffith

University I received a personal letter from the Vice-Chancellor, Roy Webb, congratulating me on being elected a member of Academia Sinica.

In August I attended the 5th International Conference on the History of Chinese Science held at the University of California, San Diego. I delivered a plenary lecture on the Ziping 子平 method of fate-calculation. Needham appeared on video delivering the congratulatory message at the opening ceremony. I was shocked to see him seemingly many years older than the time we last met in the previous year. Many of his friends present at the conference gained the same impression. It suddenly dawned on me that I would be needed in Cambridge before too long. Being a new member of Academia Sinica I received special attention from the conference delegates who came from Taiwan. I reciprocated and came across an excellent presentation by a young astrophysicist, Dr Huang Yi-long 黄一农. To confirm my impression, I got Nathan Sivin to join me in listening to his presentation. I found an up-and-coming historian of Chinese science.

Sure enough, on my return to Brisbane letters kept coming from several quarters urging me to give a definite date of taking up duty in Cambridge. During the university vacation I paid a short visit to Cambridge to get myself familiarised with the work of the Needham Research Institute. Needham put me up in a fellow's guest room at Caius College, and we went up together to dine at the College Hall taking the kitchen's lift. He was finding it difficult to walk. I decided that I ought to consume all my earned leave before resigning from my post in Griffith University, taking the opportunity to explore potential sources of funding for the Needham Research Institute. I had already discovered during my visit to Cambridge that the Institute could not be totally dependent on the New York and the Hong Kong Trusts.

Dr Wu Ta-You, President of Academia Sinica invited me to visit the academy. The three-month visit would be hosted by the Institute of Mathematics, the Institute of History and Philology, and the Institute of Modern History. My assignment was to make a survey and report to him on the state of the teaching and research of the history of science in the universities in Taiwan. I left Brisbane on New Year's Day of 1989 for Singapore and arrived in Taipei on 2 January. I proceeded to the Academia Sinica in Nankang and checked in at the Academic Activities Centre. Dr

Lih Kuo-wei 李国伟, Director of the Institute of Mathematics, was my main host, but because all the offices in his Institute were fully occupied, I had one office at the Institute of History and Philology, and another at the Institute of Modern History. I gave another talk on Needham at the Institute of Modern History.[109] Every year the Institute of History and Philology organised jointly with the Taiwan National University a memorial lecture in honour of Dr Fu Ssu-nien, who was a past director of the Institute as well as President of the National Taiwan University. The organisers invited me to deliver the 1989 lecture.[110] I made several visits to the Tsinghua University at Hsinchu, where Li Yih-yuan 李亦园 was Dean of the College of Humanistic and Social Sciences, and Chang Yung-tang 张永堂 was the Director of the Institute of History. I gave a talk and held discussions at the Institute of History, while Li Yih-yuan informed me about the impending formation of the Chiang Ching-kuo 蒋经国 Foundation for International Cultural Exchanges. He was soon to become the Executive-President of the Foundation and he wished to enlist my help. We spoke about the ways the Needham Research Institute could receive grants from the Foundation. Among the other places where I gave lectures were the National Library in Taipei, the Tunghai University in Taichung and the National Chengkung University in Tainan. I was also the plenary lecturer for the 2nd Symposium on the History of Chinese Science.[111] I discussed the fundraising problem with Dr Hsiung Ping-chen 熊秉真, a researcher at the Institute of Modern History. She helped me to contact the Representative of the Glaxco

---

[109] See Ho, P.Y. (1989), "*Zai tan wo dui Li Yue-se he Zhongguo kexue jishu shi de renshi* 再谈我对李约瑟和中国科学技术史的认识" (Speaking further on Joseph Needham and history of science as known to me, in Chinese), *Newsletter for Modern Chinese History*, 7: 25–34.

[110] See Ho, P.Y. (1989), "*Yishu yu zhuantong kexue de guanxi* 易数与传统科学的关系" (The relationship between I-numerology and traditional science, in Chinese), *Bulletin of the Institute of History and Philology, Academia Sinica*, 60.3:493–505.

[111] See Ho, P.Y. (1989), "*Cong kejishi guandian tan zhuantong sixiang zhong de shu* 从科技史观点谈传统思想中的数" (On the traditional concept of *shu* from the viewpoint of the history of science and technology, in Chinese), *Papers on Chinese Studies*, 3:1–16.

Foundation in Taiwan, requesting for the endowment of a chair in Cambridge for the Needham Research Institute, mentioning the interest of the Institute in Chinese pharmacology. The Representative was very sympathetic, but said he had no power to take action on a request from an institution outside Taiwan. He then transmitted our request to Britain for action. However, no further news was heard from Glaxco. After submitting a report on my survey of the status of teaching and research on the history of Chinese science in Taiwan to President Wu Ta-You, I returned to Brisbane on 4 April.

In early August I attended a conference in Taipei organised by the National Taiwan University. I took the opportunity to tell my Taiwanese audience something more about the work of Needham.[112] I then gave a talk at the National Tsinghua University and met Li Yih-yuan to explore the possibility of getting grants for Cambridge. In September I received a telephone call from Needham announcing his marriage to Lu Gwei-Djen.

In the same year I was appointed Honorary Professor of the Chinese Academy of Science. The Vice-Chancellor of Griffith University wrote me a congratulatory message. I asked my university for permission to go on early retirement at the end of November 1989 so that I could make preparations to go to Cambridge. My service at Griffith University ended in November. The university conferred on me the title of Professor Emeritus.

Preparations for Cambridge

I was aware of the financial problems faced by the Needham Research Institute at that time. It received no financial support from the government and also none from the university, being an independent entity. The Hong Kong Trust was responsible for helping with the building fund, while the New York Trust raised money to support the writing of some of the volumes of *Science and Civilisation in China*. In an effort to fill the gap, I decided to make

---

[112] See Ho, P.Y. (1992), "*Li Yuese yu Zhongguo kejishi* 李约瑟与中国科技史 (Joseph Needham and the History of Science and Technology in China, in Chinese), in *Minguo yilai guoshi yanjiu de huigu yu zhanwang yantaohui lunwenji* 民国以来国史研究的回顾与展望研讨会论文集, Taipei, pp. 7–25.

a private trip in December to Singapore, Hong Kong, Taiwan and Japan. I came to realise fully the wisdom in the popular saying "man proposes, God disposes", with the Chinese equivalent *"moushi zai ren chengshi zai tian"*, in matters relating to Needham. I learned not to take things personally and not to put blame on whosoever or whatsoever. That does not mean that my fundraising efforts were without success. In fact there were even isolated instances of success with little effort. Nevertheless, my December trip reminded me of the line "Failure is the mother of success" that I learned from my transcription exercise book in school.

In Singapore Tai Yu-lin 台玉玲 told me that she had made an appointment for me to see an elderly banker in his office to consider a request from me on behalf of the Needham Research Institute. I went to the meeting taking along a list of requests prepared by Cambridge with high hopes of success, because Tai Yu-lin knew the banker well and would not have made the appointment for me if she were not optimistic about the outcome. My appointment was at noon. That morning I read on the front page of the newspaper about the banker's domestic problems. He was not in a mood to talk when I saw him. He took my letter, but did not read it in front of me. It was sheer bad timing.

In Hong Kong Philip Mao had earlier introduced to me a banker who had close business connections with the Fuji Bank of Japan. I had discussed with him my intention to write to the Fuji Bank to ask for a donation for the Needham Research Institute, and asked whether he could transmit my request and put in some kind words. After his agreement to do so, I went to Tokyo to see Shuntaro Ito 伊东俊大郎 to find out more about the CEO of the Fuji Bank, noting that they were both graduates of Tokyo University. Knowing the importance of university affiliation in Japanese society, I discussed my letter with Ito and included his name as reference. Ito was then Professor of History of Science at Tokyo University and a member of the Japan Prize board. I went to Hong Kong to hand my letter of request personally to the banker. The letter was written in Japanese. Nothing further was heard of it. Ito later informed me that the Fuji Bank had not contacted him either. I did not know where my letter finally landed. When I was in Tokyo Shigeru Nakayama also accompanied me to see the Toyota Foundation, but we found that the Foundation was not directly accessible to the Needham Research Institute. I had earlier learned the importance of personal connections, without which sending letters of appeal would not achieve much. I had come to realise that, even with personal connections, the Chinese equivalent to the saying "man proposes, God disposes" still mattered, but that should not deter one's efforts.

In Taiwan I discussed with the relevant authorities the sending of delegates from Taiwan to participate in the 6th International Congress on the History of Chinese Science to be held in Cambridge in 1990 to honour Needham's 90th year. I was invited to give a talk at the National Tsinghua University at Hsinchu. Li Yih-yuan, the Dean of the College of Humanistic and Social Studies, had become the Executive-President of the Chiang Ching-kuo Foundation for International Cultural Exchanges. He asked me in my capacity as a Member of Academia Sinica to be a member of his review boards. I was then the only Member of Academia Sinica in both Britain and Australia. As a member of the review board I would be disqualified automatically from receiving any grant from the Foundation and from submitting any applications for a grant on behalf of the institution where I was employed or for any third party. As I could not sacrifice the interests of the Needham Research Institute, Li Yih-yuan said that since I was not going to be employed in Cambridge in the true sense of the word, he would accept applications from the Needham Research Institute provided they were not submitted through me. That provided me with the opportunity to offer my services for the promotion of Chinese studies in Europe and the Pacific area, besides serving the Needham Research Institute. I believed that one should never remain at the receiving end, but should sometimes take the opposite end instead. From 1990 until 2001 I served on the review boards of the Chiang Ching-kuo Foundation and witnessed many successful applications for grants from Cambridge during the 12 years.[113]

---

[113] I abstained from participation in voting in all the cases involving the Needham Research Institute.

7

The Needham Research Institute

Difficult Years in Cambridge

I took over from Needham as Director of the Needham Research Institute with my eyes wide open. From my last two visits to Cambridge and from the correspondence received from different interested parties, I had a good picture of the complexity of the situation. Fortunately for me it was not an employment in the true legal sense of the word. I need only to give my level best to this new appointment. I was prepared to step down when I felt that it would serve no purpose for me to carry on or as soon as the Institute found the means to employ a paid director. Philip Mao was fully aware of my intention; the mutual trust between us gave me some moral support to carry on.

It is not the intention here to give a detailed account of the complex situation in Cambridge at that time, nor is it to apportion blame on any particular party. Needham was in his 90th year when I took over the directorship of the Institute from him. He could barely stand without the help of a walking stick, and was generally confined to his wheelchair. He was suffering from Parkinson's disease and osteoarthritis, but maintained his presence in his office, unlike Lu Gwei-Djen, who only dropped in at her office once in a while. His productivity in bringing out new volumes in the *Science and Civilisation in China* series was greatly affected by his absence from Cambridge to raise funds for the Needham Research Institute during the better part of the 1980s. When the New York Trust raised funds to support the *Science and Civilisation* project, it appeared that the donors were promised certain dates when the volumes would be published, but these had fallen behind schedule. Needham emphatically denied that he had ever made such a promise; the conclusion was that someone not involved in the writing of the *Science and Civilisation in China* series had made the promise on his behalf. Such an excuse was not acceptable to New York. The Trust in Cambridge tried to remedy the situation by taking as much work off Needham as possible, giving him more time to attend to writing. The secretary, for example, would open all his letters, except those that were personal. This futile attempt sounded good only in theory.

Those who were not directly involved in the history of Chinese science would not have known that while Needham was moving his books and office first from Caius College to a pre-fabricated warehouse at Shaftesbury Road, near the Cambridge University Press, and then to another building at Brooklands Avenue near the Cambridge Railways Station during the latter part of the 1970s, and his shuttling between Cambridge and East Asia from 1979 to 1986, much water had flowed under the bridge in the study of history of science. There was growing emphasis on the social, political, religious and economic aspects of the history of East Asian science, particularly in the field of Chinese medicine. This new development, which also helped to bridge the gap between science and the humanities, became an important direction in the future of East Asian history of science studies. A typical example was the study of the history of Chinese medicine in a social science context, which had become rather trendy among visiting scholars from Europe and America to the Needham Research Institute. The popularity of Chinese medicine coincided with the time when Needham was occupied with the drafting of the medical section of *Science and Civilisation in China.* Vast amount of research publications from Beijing and the West had steered towards the social aspect of Chinese medicine. Needham had spoken to me about the difficulties he encountered trying to catch up with all the new literature on the history of Chinese medicine. Besides, what amount of work could one expect from an infirmed elderly scholar approaching the age of 90? Rightly or wrongly, the intended solution of the problem also gave rise to a suspicion of attempts to usurp power from the founder. Hong Kong's thought was entirely for Needham, thinking that everyone concerned was there to serve him.

In retrospect, given the position of the Needham Research Institute at that time, I doubt that a paid director could carry on for long. I myself felt like one who volunteered to drive a motor car with a full load of passengers and discovered that not only were there back-seat drivers among my passengers, but there were also different instructions coming through the mobile telephone from two well-wishers who were not travelling in the car. However good the intentions of my back-seat drivers and well-wishers were, I was finding it impossible to drive safely. Fortunately, I was prepared to vacate the driver's seat to let my passengers or the well-wishers take over. On the other hand, an employed chauffeur would have to think twice before quitting.

I took up my appointment as Director of the Needham Research Institute on 23 February 1990. Through a letter from Needham, the Gonville and Caius College made me a member of the Senior Combination Room, and through a letter from Michael Loewe, Robinson College made me a senior member of the College, giving me dining rights at both colleges. Needham soon broke the news to me that he had asked Huang Hsing-tsung to be the project coordinator of *Science and Civilisation in China*, and that the latter had already accepted the job without a salary. This could have been interpreted as a sign of a lack of confidence in me in favour of Huang Hsing-tsung. However, when I retired officially from employment in 1989, I felt that I have already had a good innings in my career and no longer saw, at my age, any reason to compete with anybody for favour, or for power. It was pointless to take that as an insult by thinking that the least Needham could have done was to consult me beforehand. The constructive thing to do was to think positively and regard Needham as having done that deliberately to relieve me from the burden of looking after the *Science and Civilisation in China* project and its associated routine problems, so that I could devote myself to the things that he had not been able to do himself in the past for one reason or another. I even told Needham that I would nominate Huang Hsing-tsung as Deputy Director of the Institute in charge of finance to give him a proper standing in the Institute. At the Trustees' meeting that approved my nomination, Dr Michael Loewe also became a Deputy Director.

Needham could not have his cake and eat it. While he was very single-mindedness in his *Science and Civilisation in China* project, for the latter part of the 1980s he was busy shuttling between Cambridge and East Asia, mainly on behalf of his Institute. His single-mindedness was a reason for his success in establishing his Institute and in the *Science and Civilisation in China* project, but it left him little time to reflect on other mundane, but essential matters. These include the careers of those working with him in his Institute, the future of his Institute after the completion of the *Science and Civilisation in China* project, ways to cooperate with the University of Cambridge, and to promote research and teaching in the history of Chinese science. For example, the Faculty of Oriental Studies of the University had approached Needham on several occasions concerning the participation of his Institute in university teaching and research. Needham was always adamant that the first priority of his Institute was his project. Negotiations produced no consensus of opinion and no solution, reminding me of a stalemate in the game of chess.

Needham's single-mindedness in his work played an important part in his *Science and Civilisation in China* project. His way of dealing with criticisms was to apply the saying "The dogs bark and the caravan moves on". One may call this arrogance, but it did give him the encouragement to carry on. His single-mindedness might have caused offence to some, but it was something that I must overlook in the overall interest of the history of Chinese science. I felt that I was no longer morally bound to the *Science and Civilisation in China* project, but that did not mean I ignored the project from then on. I was always there when the person looking after the project or a collaborator came for my advice. The freedom I gained enabled me to participate in a wider circle of academic activities in Cambridge rather than being confined to the precinct of the Needham Research Institute itself. For example, I became a member of the Internal Committee of the East Asian Studies Centre of the University of Cambridge, and I attended meetings of the Faculty of Oriental Studies Board. I received many publications from research institutes in Taiwan in my capacity as a member of Academia Sinica. The Cambridge University Library, the Library of the Faculty of Oriental Studies and the East Asian History of Science Library of the Needham Research Institute were happy to take over those books from me with my compliments. Giving books away in Cambridge was a happy experience in contrast with another place that shall remain unnamed. In Cambridge the librarian in charge of the Oriental Collection came personally to collect my books with a smile; it was an occasion when both donor and receiver were happy. In another place a junior member of the library staff came to collect the books, giving the impression that I was giving more work to that library. It is not always easy to be a donor.

About a fortnight after Needham told me about Huang Hsing-tsung, one of his collaborators in the *Science and Civilisation in China* project came to my room. After introducing himself he said that Needham sent him to see me so that I could apply to the Chiang Ching-kuo Foundation for a grant to enable him to carry on. I explained to him my position, without having to go and see Needham. I was grateful that Needham had spared me from the job of project coordinator. Huang Hsing-tsung did not seem to relish his work as project coordinator either. Somehow the need for a project coordinator disappeared when Dr Michael Loewe, the other Deputy Director, became editor. Huang Hsing-tsung remained as Deputy Director looking after finance.

I was fully aware that I had to keep up with my research to maintain my credibility as a scholar in Cambridge. Having been relieved of the burden of looking after the *Science and Civilisation in China* project, I felt free to pursue my own course of research, but at the same time I should not tread on the areas covered by the collaborators of the project. I was glad that I had already found a new avenue in the Chinese divinatory art when I was in Hong Kong. I decided to venture into an area hitherto overlooked by both Chinese scholars and western Sinologists alike, and which was outside the coverage of *Science and Civilisation in China*. In science and technology we think of something that enables us to understand or to explain nature, and the way to exploit or harness it. Modern scholars look at traditional Chinese science from their understanding of the meaning of science in the modern world. I tried to look at it from the viewpoint of the traditional Chinese, to find out what they considered that would enable them to understand or to explain nature, and the way to exploit or harness it. The term for science was absent, but the concept was there. The Song neo-Confucianists received much attention from the historians of science. Applying the above criteria for science, we can at best only regard the great exponents of the neo-Confucian school, such as Zhu Xi 朱熹, Zhang Zai 张载 and Shao Yong 邵雍 as philosophers of science and not scientists. They have never been known for their knowledge of harnessing nature. The Chinese had long been aware of the three arcane arts that were purported to enable one to communicate with nature and to predict it, such as predicting rain, snow or hail, if not to harness it. These were the three cosmic-board arts, the *sanshi* of Song China, namely *Taiyi* 太乙, *Dunjia* 遁甲 and *Liuren* 六壬, which were mentioned in Shen Gua's 沈括 *Mengxi bitan* 梦溪笔谈. Also in *The Romance of The Three Kingdoms* Zhuge Liang 诸葛亮 is said to have employed one of these arts to change the direction of the wind. In our modern context these arts come under the domain of magic if not pseudo-science, hence modern scholars might have dismissed them as such. However, to the Chinese of the past, these constituted knowledge about nature and the methods to harness it. Hence they deserved attention from those interested in the history of East Asian science. My research during the later period of my time at the Needham Research Institute was directed towards this aspect.

One of the first things I did was to contact the Lee Foundation in Singapore regarding the £5,000 it promised to contribute to the Needham Research Institute in 1984. The contributions soon arrived. The amount gradually increased until it reached £30,000 a year at the time of writing.

Through the effort of Dr Christopher Cullen, the 6th International Conference on the History of Chinese Science took place in Robinson College. This conference saw the formation of the International Society for the History of East Asian Science, Technology and Medicine, with Nathan Sivin as President and Christopher Cullen as Honorary Secretary. Together with Dr Michael Loewe, the Deputy Director, we succeeded in getting a grant from the Anglo-Daiwa Foundation to support two visiting Japanese scholars to the Needham Research Institute. I gave a lecture at the School of Oriental and African Studies, and was appointed Professorial Research Associate of that School.[114] I served the University of London on several occasions either

---

[114] See Ho, P.Y. (1991), "Chinese science: the traditional Chinese view", *Bulletin of the School of Oriental and African Studies*, 65.3:506–519.

as an external examiner or an internal examiner in examinations for Ph.D. candidates.

On the appointment of Lu Gwei-Djen as Honorary Professor of the Chinese Academy of Science, the Institute of the History of the Natural Sciences in Beijing sent someone to Cambridge to interview her to write her biography in order to introduce her to the Chinese audience, but Lu Gwei-Djen refused to be interviewed and asked the interviewer to refer to the *Who's Who*. Chen Meidong 陈美东, the Director of that institute, appealed to me. I said that I was too close to her to write her biography, but made a compromise by writing an introductory note to meet the requirement of his institute.[115] That publication pleased Lu Gwei-Djen no end. She liked it so much that she read aloud to Needham a remark I made about her. The remark said that without a Lu Gwei-Djen, there would not have been a Joseph Needham to write *Science and Civilisation in China*. She asked me for an extra copy of the reprint to give to her friend Mrs Lianli Jackson (maiden name Dai Lianli 戴连理). In their last years Lu Gwei-Djen often went with Needham to visit Lianli Jackson at her home during the weekend for an evening meal. I joined them on a few occasions. Sometimes Lianli Jackson would join Lu Gwei-Djen for lunch at Robinson College. Lianli Jackson was the widow of Meredith Jackson, former Professor of Law at Cambridge University and teacher of Lee Kuan Yew, the founding father of the Republic of Singapore. Lucy and I had been a guest of the Jacksons in the late 1950s. Lianli Jackson used to tell us that in spite of his social status, the Prime Minister of Singapore often called on his teacher during his visits to Cambridge. She died in early 1991, and Lu Gwei-Djen said that she had lost her only friend in England. By this she meant someone from the same generation. Previously she told me that she had only two friends. One was Mrs Platt, who returned from Shanghai with Professor Benjamin Platt to live in London. In the early 1930s Lu Gwei-Djen worked in the Lester Institute in Shanghai, where Benjamin Platt was the Director. Lu Gwei-Djen

---

[115] See Ho, P.Y. (1990), "*Lu Guizhen boshi jianjie* 鲁桂珍博士简介 " (A Brief Introduction of Dr Lu Gwei-Djen, in Chinese) *China Historical Materials of Science and Technology*, 11.4:25–27.

used to travel to London to spend the weekend with the Platts until the death of Mrs Platt.

In the last week of June 1990 Lucy joined me on my trip to Taiwan to attend my first Academia Sinica meeting at Nankang. Academia Sinica holds a meeting of members once every two years. In those days it provided members with the air tickets to travel to and from Taipei for the meeting. The Chiang Ching-kuo Foundation holds its review boards' meeting every year in Taipei and provides members with the air tickets to travel to and from the meeting. I took advantage of these free tickets to stop over in various places to carry out my duties as Director of the Needham Research Institute. This time we went by way of North America in order to meet the Chairman and members of The East Asian History of Science, Inc. in New York. I understood that since 1987 there were problems between The East Asian History of Science Trust and the two Trusts in Hong Kong and New York over requests from Cambridge on funding. Hong Kong's stand was that its responsibility was to raise money for the building, while that of New York was to help fund some volumes of *Science and Civilisation in China*. The two Trusts asked Cambridge to make efforts to raise funds in the UK and Europe to meet its own needs instead of depending on them. The New York Trust also queried the slow progress of the production of new volumes in the *Science and Civilisation in China* series. I could not say that the series was not my responsibility so I had to play dumb. The meeting got nothing out of me to commit myself to actions that might upset Needham. An unpaid piper could not be asked to play a tune, and, in this case, to play a tune against his will.

We went on to San Francisco, where Lucy met her eldest sister Fanny (冯范如) and her husband Kenneth (杨炯暄) and the rest of the family. Leaving San Francisco we went to Taipei to attend my first meeting at the Academia Sinica. The Institute of Modern History asked me to report on Needham and his Research Institute.[116] Our next stop was Hong Kong,

---

[116] See Ho, P.Y. (1991), "*Li Yuese yanjiusuo yu Zhongguo kejishi yanjiu*" (The Needham Research Institute and its research on the history of Chinese science, in Chinese), *Newsletter for Modern Chinese History*, 12:45–49.

where we met with Mr P.L. Lam 林炳良 and Mrs Mary Lam, both old friends of Needham, to talk about financial contributions to the Needham Research Institute. We stopped over in Singapore to see our son Yik Hong and Lucy's sister Annie (冯仲姬) on our way to Brisbane.

After a short stay at home in Brisbane I rushed back to Cambridge. On 18 August I left Cambridge again, to go to Taiwan and Beijing. I stopped over briefly in Kuala Lumpur to see my infirm mother. That was the last time I saw her; she passed away a few months later while I was travelling abroad. On 23 August I left Kuala Lumpur for Taiwan, where I attended the International Conference on the History of Science and Technology in Modern China held at the National Tsinghua University in Hsinchu. I presented a paper concerning the spread of western science in early 19th-century China[117]. Through Hsiung Ping-chen I became one of the directors of the "Centre for Medicine and Culture in China", which was sponsored by Glaxco Taiwan Ltd. I was hoping that this connection would be of some help to the Needham Research Institute one day. However, I have heard nothing about the Centre since the time I witnessed its formation.

On 29 August I left Taipei for Hong Kong and stayed at the Robert Black College. That evening I had dinner with the Vice-Chancellor of Hong Kong University, Wang Gungwu, at his Lodge. Wang Gungwu said that he would be hosting the 34th International Congress of Asian and North African Studies in August 1993 and asked for my help in organising a special session on Asian science with reference to inter-cultural contact in history. I was happy to comply with his request. The next day I had lunch together with P.L. Lam and Mary Lam. On 3 September I attended the Hong Kong Trustees' meeting and discussed the way to forward the donation of Mr K.P. Tin 田家炳 to Cambridge for the building of the South Wing. The money was raised through George Hicks, whose father-in-law was a friend of the donor.

---

[117] See Ho, P.Y. (1991), *"Cong 'Jinghuayuan' shitan shijiushiji chuqi keji zai yiban shiren zhong de puji"* (On the popularity of science and technology among scholars in general judging from the novel *Jinghuayuan*, in Chinese), *Zhongguo kejishi zhuankan* (Taipei), pp. 19–31.

I left for Beijing on 4 September 1990. Upon my arrival at the Institute for the History of Natural Sciences I found some representatives from the University of Science and Technology, Beijing, waiting for me. They took me to their university for the award of the title Honorary Professor of that university. On 5 September, at the opening ceremony of a conference on the history of Chinese science, the chairman handed me the official diploma that certified me as an Honorary Professor of the Chinese Academy of Science, and also conferred on me honorary membership of the Chinese Society of the History of Science. That was followed by my plenary lecture entitled "Looking at the history of Chinese science from the viewpoint of a traditional Chinese scholar".[118] That evening my lecture was telecast briefly on Chinese television. On 7 September the publisher of the new Chinese translations of *Science and Civilisations in* China launched a ceremony to introduce the first three volumes. I sat next to Dr Qian Sanqiang 钱三强 and spoke during the ceremony thanking the Chinese people for their help and encouragement on behalf of Needham. I received tumultuous applause from the audience, and that evening my speech was again telecast on Chinese television.

On 9 September I was given an audience by the General-Secretary Jiang Zemin 江泽民 at Zhong Nanhai. The interview scheduled just before Jiang's next appointment to receive a visit from Henry Kissinger lasted 37 minutes. It was carried out in a relaxed manner. We talked about the rapid advances in technology in our modern age, and Jiang said that he still remembered the days when he learned about the vacuum valve. When asked about my views on the Chinese economy I said that I was expecting to see the day when Shanghai would overtake Hong Kong, explaining that Hong Kong owed much of its economic success to the businessmen from Shanghai and the Zhejiang province. When I remarked that Shanghai and Zhejiang produced great businessmen, someone said in a soft voice that Jiang came from

---

[118] See Ho, P.Y. (1991), "*Shi cong lingyi guandian tantao Zhongguo quantong keji de fazhan* 试从另一观点探讨中国传统科技的发展" (Looking at the development of traditional Chinese science from another angle, in Chinese), Plenary Paper at the International Conference in History of Chinese Science held in Beijing, September 1990, *Exploration of Nature*, (Beijing) 10:27–32.

Shanghai. I looked at Jiang and apologised, saying that I would not have made that statement had I known that he was from Shanghai. I noticed his attempt to conceal a smile on his face. Finally, our conversation turned to the history of Chinese science. After expressing his support for the Needham Research Institute, he looked towards the direction of Chen Meidong, the Director of the Institute of the History of Natural Sciences, and said that one must guard against being chauvinistic when doing research on the subject. That evening Chinese television telecast my audience with Jiang.

The next day I gave a seminar at the Beijing Normal University at the invitation of Professor Bai Shangshu 白尚恕, one of the top-ranking historians of Chinese mathematics in the last few decades of the 20th century. I first met Bai Shangshu in Hong Kong in 1983 together with Professor Li Di 李迪 and Li Jimin. The trio formed a friendly group of historians of Chinese mathematics who were well-known among their peers.

On 11 September I left Beijing for Tokyo. That evening I was invited to a seminar-cum-dinner session at Shinjuku by the Society of the History of Japanese Mathematics. I talked about my lecture that was delivered in Beijing and about the activities of the Needham Research Institute. I then went to Osaka to talk to the organisers of a proposed Institute or Centre for the history of East Asian science at the Kansai Science City on matters concerning the future cooperation between the Needham Research Institute and the new centre of learning. The spokesman for the proposed project was Professor Kunio Goto 後藤邦夫. We discussed various matters of mutual co-operation between Cambridge and Kansai, including staff exchange, English translations of Japanese works, training of Japanese research students, library facilities, workshops and conferences. I returned to Tokyo on 20 September. The next morning I paid a courtesy call to the National Institute for Research Advancement (NIRA) and met its President, Dr Atsushi Shimokobe 下河边淳, the Director of the Research Planning Department, Mr Hideaki Yamakawa 山川英明, and Miss Yukiko Hirakawa 平川幸子, who dealt directly with the applications put up by the Needham Research Institute for funds. After receiving the first grant from NIRA, the Needham Research Institute had been experiencing difficulties for some time to get the second grant released. It arose from a complete misunderstanding. Although

NIRA was prepared to release the funds, the paper work done by Cambridge to justify its needs was not according to the requirement of NIRA. Dr Shimokobe instructed Miss Hirakawa to work closely with me to put the application from Cambridge in an acceptable form for the release of funds. After doing our homework, Miss Hirakawa remarked that it was a relief for her to work with me instead of dealing with a highly respected senior scholar. Actually, she had not been dealing directly with Needham himself, but with a senior gentleman who had no knowledge of the East Asian people and culture. The way for the approval of our application was cleared; I saw no necessity to give an explanation that might complicate matters. Dr Shimokobe retired the following year; his successor, Shinyasu Hoshino, assumed duty on 16 November 1991.

I left Tokyo for Singapore on the evening of 21 September. I visited the National University of Singapore, where Lam Lay Yong invited me to lunch to meet the Dean of Arts, Professor Edwin Thumboo, and the Professor of Chinese, Lim Chee Then 林徐典, all of whom were my friends. They asked me to give a series of lectures at their university at some future date. The National University of Singapore employed a full-time officer to attend to fundraising matters. He also happened to be a friend of mine. We had a good chat on the methods of fundraising, and he said that sending letters of appeal seldom produced results. I left Singapore on 26 September and arrived the next day in Cambridge.

I returned to Cambridge to see the disappearance of many of the problems that had plagued the Needham Research Institute in the past. The donation of K. P. Tin from Hong Kong enabled the completion of the South Wing, NIRA was about to approve the application for the second grant towards volume 7 of *Science and Civilisation in China*, while the Chiang Ching-kuo Foundation had given grants not only for the writing of Needham's medical volume but also for the library. The Lee Foundation of Singapore had begun making its annual contribution to the Institute. The Anglo-Daiwa Foundation had enabled a Japanese scholar to work in the Institute for a doctoral degree at the University of London. Book donations came from the direction of South Korea. The problem that was still unsettled was that from across the Atlantic concerning the publication of *Science and*

Civilisation in China. I thanked my lucky stars that Needham had unwittingly let me off the hook. The *Science and Civilisation in China* project remained the major issue confronting the Institute in the early 1990s. In December I returned home to Brisbane to join my family.

On 19 to 20 January 1991 I was in Kyoto attending the East Asian History of Science and Workshop session of the International Institute for Advanced Studies in Kansai. I gave a report on the activities of the Needham Research Institute, and participated in the discussion on the second day concerning the future role played by a proposed research centre for the history of East Asian Science at the Institute in Kansai. The Workshop reached a consensus that the proposed research centre would seek close cooperation with the Needham Research Institute in the areas of translations of Japanese books on the history of science, working facilities for Japanese scholars visiting Cambridge, cooperation between libraries, with all the expenses to be borne by Kansai. Following the Workshop the convenor, Professor Kunio Goto, attended the opening ceremony of the South Wing of the Needham Research Institute on 10 May and discussed some of the points I raised with Michael Loewe at the Institute. Goto then went on sabbatical. During his absence the plans for the research centre at Kansai could not get off the ground due to the bursting of the economic bubble in Japan.

From February to June 1991 I served the National Tsinghua University, Hsinchu, as a Foundation Distinguished Visiting Professor and conducted a course on the history of science for M.A. students in the Institute of History. My lectures resulted in the publication of a book that I dedicated to my teacher, Sir Alexander Oppenheim.[119] I wrote to the Minister of Education suggesting that the history of science be accepted as a topic for Ministry of Education scholarship holders to pursue higher degrees in overseas universities. That resulted in the first scholarship holder from that ministry to be enrolled as a Ph.D. candidate at the London School of Oriental and African Studies.

---

[119] See Ho, P.Y. (1994) *"Hai na baichuan*: *Zhongxi kejijiaoliu shi* 海纳百川: 中西 科技交流史 "* (The Sea Admits the Hundred Streams: History of transmission of science and technology between China and the West, in Chinese), Taipei.

When I stopped over in Hong Kong I always made it a point to visit P. L. Lam and Mary Lam, as they were devoted and old friends of Needham who had contributed much financial help to Needham and his Institute. The Lams told me about their plans to organise a display of Chinese artworks in ivory, depicting various events in Chinese history, mainly to attract tourists to Hong Kong. They made a promise to me that part of the admission fees collected would go to the Needham Research Institute as income. They asked Needham to contribute a congratulatory message for the opening of the display, and I was to give a Chinese translation of that message, besides reading it at the ceremony.[120]

On my return to Cambridge from Taiwan I served as president of the History of Science section for the British Association of Science in 1991 and attended its annual conference in Plymouth. I also gave a lecture at the University of Durham and spoke to the school leaving class of a girls' school in Newcastle on the Chinese language. The future of the Institute should be looking much brighter than a year ago, but internal problems in the Institute developed during my absence in East Asia. The second half of 1991 was an uncomfortable period for the Needham Research Institute. Originally, I had accepted an invitation from the University of Malaya to go to Kuala Lumpur as a visiting professor for a short period. That would have provided a good opportunity for me to renew my acquaintance with the Malaysian society. I was asked by Hong Kong to take that opportunity to approach a wealthy Chinese, one Mr Lim from Penang, whom I did not know, for a sizeable donation.[121] I would be the houseguest of the Head of State of Penang when I visited that island, and would try to contact Mr Lim through my host. However, internal problems in the administration of the Institute forced me to abandon my trip to Malaysia. One of my two Deputy Directors had resigned, followed later by the resignation of the Chairman of the Trustees for some other reason. Needham and Lu Gwei-Djen were very concerned about the replacements of the two vacancies, particularly that of the new Chairman. Lu Gwei-Djen's health also began to deteriorate. In November she suffered from

---

[120]Unfortunately, due to a ban on the use of ivory in artworks, P.L. Lam decided not to launch the display.

[121]I cannot remember his name now. In any case he has since passed away.

a bad cough. Antibiotics did not seem to have much effect. Her doctor and nurses came regularly to attend to her. In the morning of Thursday, 21 November 1991 I bade farewell to Needham and Lu Gwei-Djen, as I had to catch a bus in Cambridge for Heathrow to board a flight to Singapore that evening. That was the last time I saw Lu Gwei-Djen.

My purpose of going to Singapore on this occasion was to accept an appointment by the National University of Singapore as the Isaac Manasseh Meyer Fellow for the period 1–15 December 1991. I arrived a little earlier with the hope of suggesting to one of the universities in Malaysia to develop a centre for the study of Islamic science. I had written earlier to my child-hood friend Hamdan, who was then the Head of State of Penang, about my intention and he was very supportive of the idea. I went to Penang via Kuala Lumpur, where I stopped over to see my brothers and a sister. In Penang I was the houseguest of the Head of State at his official residence, Seri Mutiara. Hamdan spoke to the Vice-Chancellor of *Universiti Sains Malaysia* over the telephone to make an appointment for my visit to the university. I received a warm welcome from the Vice-Chancellor, who told me of the interest of his university in Islamic science. He asked the staff member in charge to meet me and left the two of us alone to discuss the subject. To my dismay I found that he was talking about religious science, and I did not think that my idea could get across. I had the feeling that campus politics would hamper the development of a centre of the history of Islamic science in that university; I should not try to push for anything that I was not personally involved in. I stopped over in Kuala Lumpur again on my way to Singapore. I called on the new Vice-Chancellor of the University of Malaya, Professor Osman bin Taib, who was my Deputy Dean when I was the Dean of Arts at that university in 1967. I spoke to him about my visit to Penang and he agreed with my assessment on campus politics. I then returned to Singapore and stayed at Temasek Hall within the campus of the National University of Singapore from 1 to 15 December. I gave a public lecture on a Chinese method of fate-calculation.[122] The news of Lu Gwei-Djen's demise came

---

[122]See Ho, P.Y. (1991), "*Suanming shi yimen kexue me?* 算命是一种科学么" (Is fate-calculation a branch of science?, in Chinese), *Xuecong*, 3:1–20.

during my stay at the campus of the National University of Singapore. I had to give interviews to the local newspapers about her passing and the welfare of Needham.

The Institute Gradually Getting Back on the Rails

I originally planned to return to Cambridge in spring, but at the request of Philip Mao, I turned up unexpectedly to attend the Trustees' meeting in Cambridge in January to elect a new Chairman. The meeting elected Professor Sir Geoffrey Lloyd, the Master of Darwin College, as Chairman and Christopher Cullen as a Deputy Director. The new Chairman found time to take a personal interest in the Institute. He attended to Institute affairs in person and participated actively in the text-reading seminars held at the Institute. These seminars were somewhat similar to those first organised in Kyoto. At a seminar the speaker would read, translate and discuss a selected Chinese text with the participation of the audience. It usually drew a good attendance from members of the Institute and the University, besides the occasional presence of scholars from Britain, Europe, North America and East Asia. Needham also made it a point to attend the seminar. His very presence was an inspiration to the many scholars from China. On the other hand, the presence of young scholars added the atmosphere of youth to the Institute. After the demise of Needham some of the seminars were conducted at the campus of Cambridge University. Christopher Cullen was in his prime, almost 25 years my junior. I could leave many things in his capable hands, without ever trying to be a back-seat driver. He looked after the *Science and Civilisation in China* project with great efficiency, and I was quite confident that he would eventually see it to completion.

I returned to Brisbane after the Trustees' meeting. In April I attended the meetings of the Chiang Ching-kuo Foundation review boards in Taipei and took part in an International Conference on the History of European Sinology at the Grand Hotel, which was organised by the Chiang Ching-kuo Foundation. In Taipei I also paid a courtesy call to Mr Chen Li-fu 陈立夫 on behalf of Needham. Chen Li-fu sought my help to negotiate with the Cambridge University Press over the copyright issue arising from the Chinese translations of *Science and Civilisation in China*. Although he was a few months older than Needham in age, Chen Li-fu was still robust and influential in those days. I stopped over in Tokyo, Hong Kong and Singapore on the way. In Hong Kong I discussed with the Hong Kong Trustees the subject of supporting a young scholar from China to do a Ph.D. degree at Cambridge University. I also discussed about Needham's state of health with P.L. Lam and Mary Lam. They had earlier contributed directly to the support of a research assistant to help Needham during his office hours at the Institute.

I returned to Cambridge on 30 April. On 2 May I attended the memorial service for Lu Gwei-Djen held at Robinson College, of which she was a Founding Fellow. I was trying hard to find a way for the Institute to bring over a promising young Chinese scholar for a year. One day a fellow member of Academia Sinica from the US, Professor Leong Way (梁栋材), visited the Institute. I spoke to him about my plan. To my surprise, he told me that I could apply to the Li Foundation of New York, of which he was the President and his wife the Honorary Treasurer, for a grant. This was shortly after Lu Gwei-Djen's memorial service; my friend suggested naming the fellowship in honour of her. That was the beginning of the Li Foundation of New York Fellowship at the Needham Research Institute.

Lucy came to join me in Cambridge to see how Needham was managing after the demise of Lu Gwei-Djen. In May 1992 Lucy and I visited Edinburgh together with Chen Meidong, the Director of the Institute for the History of the Natural Sciences in Beijing, who was on a visit to Cambridge as a guest of the Needham Research Institute. I was looking forward to building closer links with that Institute by inviting Chen Meidong and Liu Dun 刘钝 to Cambridge. I delivered a public lecture on the *Book of Changes* and demonstrated its method of divination using chopsticks at the University of Edinburgh. I hosted a dinner at one of the popular Chinese restaurants and while talking to its proprietor I found out that he was a leader in the local Chinese community. I told him that the Chinese community should take a greater interest in the university.

On 20 June 1992 Lucy and I left London for Tokyo, arriving on 21 June. Shigeru Nakayama invited us to stay at his apartment at Yoyogi-

Uehara. On 24 June we went to Osaka to talk to Professor Goto about the possibility of seeking funds from Japan to finance some joint projects between Kansai and the Needham Research Institute. I found that Goto was no longer optimistic. I gave a talk at the National Seoul University in Seoul on 2nd July and paid a courtesy call to the Daewoo Foundation to thank the President for his donation of Korean books to the East Asian History of Science Library of the Needham Research Institute. We left Seoul on 4 July for Taipei where I attended the Academia Sinica members' meeting for the year. We met Professor Leong Way and his wife, who became great friends with Lucy, claiming her as a sister.

Lucy and I then returned to Brisbane via Singapore. After a brief stay at home in Brisbane, I had to go to Singapore again, this time as the Tan Kah Kee Foundation Forum Lecturer for 1992. Through contacts established by me, this Foundation later played host to invited delegates from China to the 1999 International Conference of the History of East Asian Science at a welcome dinner. In August 1992 I attended the 7th International Conference of History of Chinese Science held in Hangzhou, staying at the Xizi Hotel by the side of the West Lake. The newspaper *Wenhuibao* 文汇报 interviewed me about Needham. I spoke of Needham who still went to his office daily in his wheelchair, and on the medical volume as well as the last volume of his monumental work.[123] (See Figure 22.) I also paid a visit to Hangzhou University to meet Professor Xue Yanzhuang 薛艳庄, past President of that University, whom I first met in 1973. I returned to Cambridge after the conference.

Things at the Institute were very much back on the rails with Geoffrey Lloyd and Christopher Cullen. The issue of publication still remained, but it was efficiently handled by youthful hands. A bigger concern was the condition of Needham, who was getting more feeble and vulnerable with age. Since I took up office in Cambridge in 1990, Needham had been occupied with his medical volume and was trying to get someone to help him to write it. One day he had a native East Asian visitor from North America, with the credentials of a medical doctor and North American residence. The visitor

---

[123] See *Wenhuibao*, 27 August 1992, page 3.

本报8月26日杭州专电（特约记者万润龙）李约瑟的学生、英国剑桥李约瑟研究中心主任何丙郁先生今天面对150多位参加中国科技史国际学术研讨会的海内外学者，代表李约瑟先生祝贺这次国际学术会议在大陆杭州召开。他详细介绍李约瑟的近况，形成了

学术会议上露面，李约瑟博士对此十分关注。在世界各国举行的国际学术会议上，李约瑟博士一旦有发言，必然提出愿望，希望中国学者早日再在国际学术舞台上活动。自八十年代以来，中国的科技史专家不仅参与了国际间的学术会议，而且在国内组织了多次国际性的学

只能用录音机口述。李先生以毕生精力研究中国的科学与文明，早在10年前他与鲁桂珍博士已着手编写《中国科学与文明》卷6"医学史"篇。今年7月，李约瑟博士宣布："医学史"篇的书稿已大功告成，将由剑桥大学出版社出版。继"医学史"篇完稿之后，目前他正提前写作《中国科学与文明》的卷7"总结"篇。他之所以提前写此卷，是考虑到"岁月催人"，李博士认为，将

何丙郁在杭介绍李约瑟博士近况

身坐轮椅 口述专著

提前写作《中国科学与文明》卷七"总结"篇

今天开幕的这次国际学术会议的一个高潮。

何丙郁先生回忆说，自1956年竺可桢先生和几位中国代表出席在意大利佛罗伦萨召开的第8届国际科学史学术会议后相当长的一段时间中，中国大陆的科学史家没有在国际

术会议，对此，身为研究中国科学与文学的李约瑟先生感到了由衷的欣慰。

据何先生介绍，李约瑟先生每天依旧上班，中午12时靠人推动他的轮椅到研究室，下午1时15分在人陪同进午餐，5时下班。因为写字有困难，他

来全部书完成时所得的结论也不会比他写的相距太远。

介绍了李约瑟博士的近况后，身为中国科学院名誉教授的何丙郁先生希望与与会专家合作，共同努力研究和发扬东亚地区的科技史。

Figure 22 Report from the *Wenhuibao*

told Needham that he knew many foundations in Japan that would be willing to provide grants to his project, and that he had found some connection between quantum mechanics and Chinese medicine. Needham took him on for a trial period and introduced him to me. That was the first and only time we saw each other, as he next came when I was away overseas. On his third visit to Cambridge he asked Needham to apply to a foundation in Japan to enable him to collaborate in writing the medical volume. Needham wrote as he requested and sent the letter through me while I was in Taiwan asking me to add my support. Being sceptical over the claim that quantum mechanics had something to do with Chinese medicine, I simply passed the letter on to its destination. Not long afterwards Needham became annoyed with the visitor, who became too familiar by addressing him as "Joe". Needham sent for Christopher Cullen, saying that he did not wish to see this gentleman again. Cullen discharged his duty very efficiently; the application for the grant was not successful either.

I remembered the ancient Chinese motto *ru jing yi wen jin, ru xiang yi wen su* that my father used to teach his pupils, to learn about the customs and the laws of the land wherever one went to. Being unfamiliar with local laws and customs, I exercised great care in matters concerning Needham's domestic problems. A Chinese doctor once told me that he was chased away by Needham's nurse when he tried to approach Needham with his stethoscope on the grounds that he was not authorised to do so, although his intention was only to find out what was wrong. That was an example of not paying heed to the motto taught by my father. Needham had appointed two of his close friends, both Fellows of Gonville and Caius College, as his executors. Brian Harland was to look after the Needham Research Institute while Ian McPherson was to take care of his personal affairs. With regard to Needham's domestic problems, I always remained friendly with Dr Ian Macpherson, and never tried to interfere; I did what I could only upon request.

The Chinese Ambassador, His Excellency Ma Yuzhen 马毓真, was very concerned over Needham's state of health. As it was difficult for Needham to go to London to have dinner at the ambassador's residence, he sent his chef to Cambridge to cook dinner for Needham. That meant dinner for everyone in the Institute. This happened more than once.

On 22 October Needham received the award of Companion of Honour from Her Majesty Queen Elizabeth at Buckingham Palace. That was also the month when John Moffett, the new Librarian, took up office at the Institute following the resignation of Dr Hilary Chung.

I remained in Cambridge until December and returned home before Christmas to be with my family in Brisbane. In early April 1993 I went to Hong Kong to help in a special session on Asian science with reference to inter-cultural contact in history at the 34th International Congress of Asian and North African Studies. While in Hong Kong, through the introduction of Rayson Huang, Louis Cha (查良镛), the famous Hong Kong novelist and newspaper proprietor, invited Professor David McMullen and me to dinner at the Mandarin Hotel. Philip Mao understood my desire to find a successor and that he or she must be paid a full salary. At his request, I wrote a letter to Dr Ma Lin at the Run Run Shaw Foundation requesting a donation towards a Joseph Needham Chair or Directorship for the Needham

Research Institute. It took a long time to find out that our efforts were in vain. Immediately after the Congress I rushed to Taipei to attend the meetings of two Chiang Ching-kuo Foundation review boards. After the meetings I went to Tokyo to make a courtesy call to the National Institute for Research Advancement together with Shigeru Nakayama. The new President of this foundation was away. We met his two Deputy Presidents, who said that in view of the economic climate in Japan it would be futile to submit new applications to the foundation for consideration in the foreseeable future. From Tokyo I returned to Cambridge.

Between 2 and 7 August I attended the 7th International Conference on the History of East Asian Science at the Keihanna city in Kansai and read a plenary paper.[124] Between 10 September and 10 December 1993 I served as Visiting Professor at the Nanyang Technological University in Singapore and conducted a course on the history of science for educationalists.

I returned to Cambridge after the New Year. In April 1994 I went to Japan to act as the internal examiner for the oral examination of a Ph.D. candidate of the University of London after attending the meetings of the Chiang Ching-kuo Foundation review boards. The candidate was a former Anglo-Daiwa Fellow at the Needham Research Institute. The venue of the examination was in Kanagawa University, at the office of the external examiner, Shigeru Nakayama. I returned to Cambridge after the oral examination. My term of directorship of five years was drawing to an end. At the request of the Chairman of the Trustees I agreed to stay on, both because of Needham's advanced age and my feeling that I would be able to work together as a team with the Chairman. Furthermore, Christopher Cullen was attending to the routine work of the Institute quite efficiently, giving me time to attend to my own research. Lucy came to join me in Cambridge in June. During this time Needham went for a vacation in Wales, accompanied by his caretaker and a friend of hers. In the last week of the month Lucy and I left together for Taipei via San Francisco, where we stayed a few days with

---

[124]See Ho, P.Y. (1995), in Hashimoto, Jami and Skar, eds., "Changing Perspectives in Historical Studies of the History of East Asian Science", *East Asian Science: Tradition and Beyond*, Osaka, pp. 7–16.

Lucy's eldest sister and her brother-in-law. After the Academia Sinica meetings at Nankang we returned to Brisbane via Hong Kong and Singapore. In Hong Kong we met P.L. Lam and Mary Lam as usual to report to them on Needham's condition. We also held discussions with Philip Mao.

After a short stay at home in Brisbane I was on my way again back to Cambridge. Normal routine work went on as usual at the Institute, but Needham's health was declining. I cannot forget the last time I was with him. On a September weekend the Needham Research Institute organised a two-day workshop to discuss the topic "Was Daoism solely responsible for the development of science in China?". The participants included the Professors of Chinese from Cambridge, Oxford and London, Sinologists from Britain and Europe, and graduate students from Cambridge. Needham was present, sitting in a corner, but the participants did not seem to notice his presence. Over the two days all those who spoke denied that Daoism was solely responsible for the development of Chinese science, which was opposite to the views adopted by Needham. Needham was physically unable to respond. I was the chairman for the last session. In my capacity as Director of the Institute I thought that I ought to say something for Needham, but as chairman of the session I ought to take a neutral stance. At the end of the session I said:

> Before I end this last session of the workshop I would like to take a little test to find out whether I can become a diplomat, with everyone of you as examiner. I shall set the question and there is no need for you to tell me my score. The question is "Show that all those who spoke were right, show also that Dr Needham was right, and show that even those who did not speak were right". I agree fully with those who spoke. There are many examples of Confucian scholars and Buddhist monks in history who made great contributions to the development of science in China. Dr Needham took the broad sense of the word Daoism, without limiting it to the Daoist religion itself. In its broad sense it included Buddhism and Confucianism. Therefore, Dr Needham was also right. Those who did not speak adopted the profound

Daoist philosophical teaching of *wuwei* 无为 — action in inaction. They knew what was right and did not see the need to speak. I now declare the session closed.

The participants left the room smiling. As he left the room, David McMullen smilingly said that I had passed the test. I replied that alas it was all too late for me. Needham called me to his side and pointed towards Tim Barrett, asking who he was. I answered that he was the Professor of Chinese History at the London School of Oriental and African Studies. That was the last conversation I had with Needham. I bade him farewell and left for Heathrow to catch a flight for Taipei via Singapore.

Between September 1994 and February 1995 I served as National Science Council Research Professor at the Academia Sinica and conducted a course on the history of science at the National Tsinghua University, Hsinchu. At the Academia Sinica I gave a seminar concerning an obscure passage in the Chinese Dynastic Histories that turned out to be a divinatory text based on the *Taiyi* method.[125] Taipei published a book on collected essays that contained an article of mine on the *Book of Changes* and Chinese science.[126] I returned to Brisbane via Singapore after my engagements in Taiwan. While I was in Brisbane I received a telephone call on the morning of 25 March 1995 from Christopher Cullen, saying that the condition of Needham was critical. He called again a few hours later to report that Needham had passed away peacefully at his residence. That was on 24 March, Cambridge time. I soon had to leave for Taipei to attend the meetings of the Chiang Ching-

---

[125] See Ho, P.Y. (1996), "*Taiyi* shushu yu *Nan Qi shu* Gaodi benjishang shichen yue zhang 太乙术数与南齐书高帝本纪上史臣曰章" (The *Taiyi* method of divination and the Historiographer's Remarks in the Chronicle of emperor Gaodi in *Nan Qi shu*, in Chinese), *The Bulletin of the Institute of History and Philology, Academia Sinica*, 67.2:383–413.

[126] See Ho, P.Y. (1995), "*Cong kejishi guandian tan yishu* 从科技史观点谈易数" (The *Yijing* system of divination from the standpoint of history of science, in Chinese), in Ho Peng Yoke *et al.*, *Zhongguo kejishi* (Collected Papers on the History of Chinese Science and Technology), Taipei, pp. 19–34.

kuo Foundation review boards. I took the opportunity to read a paper at a conference on the history of Chinese science.[127] I was then nursing a pain in my legs at home. Nevertheless I went to Cambridge when told about the memorial service for Needham, which took place at the Great St. Mary Church in Cambridge on 10 June. On that day Geoffrey Lloyd and I looked after two different groups of invited guests at lunch before the service. After the ceremony at Great St. Mary, I spoke at the Institute on its future role within the framework of the contemporary and future trend of East Asian history of science studies. I soon received a request from Oxford to write a biography of Needham. I declined on the grounds that I was too close to Needham and recommended another writer. Actually it would be an impossible task for anyone at that moment to write a full biography of Needham. It was Needham's wish that his personal files remain inaccessible until 50 years after his passing, and those files are in the archive of the Cambridge University Library. His academic files are kept mainly at the Needham Research Institute. Needham had a habit of not throwing anything away, even his parking tickets. I once mused to John Moffett that I pitied Needham's biographer, who would have to spend a lifetime going through his papers.

On 14 July I received an Honorary Doctor of Letters Degree from the University of Edinburgh. During the ceremony I walked with the aid of a walking stick that was formerly used by Needham. I travelled to and from Edinburgh by train, accompanied by John Moffett. On my return from Edinburgh I consulted an orthopaedic surgeon at Evelyn Hospital, Cambridge, on the condition of my knees. He diagnosed my pain as due to osteoarthritis, but that I needed no operation procedure at that stage.

On 29 July 1995, with the help of my son Yik Hong and my daughter-in-law Ludmilla, I attended the launching ceremony of Tai Yu-lin's *Memories of the Plague Fighter Dr Wu Lien-teh* at the National University of Singapore and received a copy of this book each on behalf of the Cambridge University

---

[127] See Ho, P.Y. (1996), "*Taiyi* shushu jiqi dui quantong kexue zhi yingxiang 太乙术数及其对传统科学之影响" (The *Taiyi* method of divination and its influence on traditional Chinese science, in Chinese), *The History of Science Newsletter*, 14:1–12.

Library, Emmanuel College and the Needham Research Institute. I walked with the aid of a walking stick. I returned to Brisbane, thinking that my travelling days might be over. My daughters changed the medication for my hypertension and encouraged me to walk as a form of exercise. The pain in my legs gradually subsided.

The Dragon's Ascent

In January 1996 I was able to attend the 8th International Conference of History of Chinese Science in Shenzhen with the aid of a walking stick. I delivered a lecture on a new approach to the study of the history of Chinese science.[128] I introduced the representative of Totem Company to Dr Lu Yongxiang 路甬祥, President of the Chinese Academy of Science, and the Director and Deputy Directors of The Institute of the History of the Natural Sciences to speak on the proposed Dragon's Ascent project. I attached great importance to this project with the hope that it would solve not only the problem of a salary for my successor but also the financial difficulties of the Institute for the History of the Natural Sciences in Beijing. I suggested what would be crucial to the success of the project was to find an influential and dedicated personality in Hong Kong who would raise the initial funds for the project. The Chinese Academy of Science together with Lady Youde, the widow of the late Governor of Hong Kong, Sir Edward Youde, succeeded in getting the help of Sir Q.W. Lee, who raised the money for the project and was closely involved in bringing it to a successful completion.[129]

Between 27 March and 6 April I was at the Academia Sinica, Nankang, to attend a symposium on the history of Chinese science. I then visited Tokyo and Kyoto to confer with Nakayama and Yano on my discovery of

---

[128] See Ho, P.Y. (1999), "*Yanjiu Zhongguo kexueshi de xin tujing: Qimen Dunjia yu kexue* 研究中国科技史的新途径奇门遁甲与科学 ", in Wang Yusheng 王渝生 *et al.*, eds., *Diqijie Gujii Zhongguo Kexueshi Huiyi Lunwenji* 第七届国际中国科学史会议论文集 (Zhengzhou), pp. 12–16.

[129] Lady Youde is presently Chairman of the Needham Research Institute Trust, renamed from the former East Asian History of Science Trust.

Hellenistic influence on a Chinese method of fate-calculation. I returned to Taipei on 11 April for the meetings of the Chiang Ching-kuo Foundation review boards. I returned to Cambridge via Singapore.

Lucy joined me in Cambridge in June. Towards the end of the month we left for Taipei via San Francisco, where we spent several days with Lucy's eldest sister Fanny and brother-in-law, Kenneth. At the Academia Sinica meeting in Nankang, Taipei, we again met with Leong Way and his wife. Lucy made several friends among the wives who accompanied their husbands to the meeting.

Between 26 and 31 August 1996 I was in Seoul attending the 8th International Conference of History of Science in East Asia at the National Seoul University. I spoke on neo-Confucianism and science at a plenary lecture.[130] After my return to Brisbane I went for a cataract operation on both my eyes and discovered that I had glaucoma as well. I spent the rest of my time recuperating at home.

In 1997 I contributed 11 articles to Helaine Selin's *Encyclopaedia of the History of Science, Technology and Medicine in Non-Western Cultures*.[131] In April 1997 I went as usual to Taipei to attend the meetings of the Chiang Ching-kuo Foundation review boards. After a brief visit to Hong Kong I returned to Cambridge via Singapore. In June 1997 I visited Berlin to advise the Technological University on the teaching and research on the history of Chinese science, and on a proposed international conference on the history of Chinese science to be held at that university in 1998. I also gave a talk on

---

[130] See Ho, P.Y. (1999), "Did Traditional Chinese Thinking Involve Using Calculations to Predict Natural Phenomena?" in Yong-sik Kim and Francesca Bray, eds., *Current Perspectives in the History of Science in East Asia*, Seoul, pp. 185–193.
[131] See Ho, P.Y. (1997), "Astrology in China", in Selin Helaine, ed., *Encyclopaedia of the History of Science, Technology and Medicine in Non-Western Cultures*, (Dordrecht) pp. 76–78; "Astronomy in China", *ibid*, pp. 108–111; "China", *ibid*, pp. 191–196; "Ge Hong", *ibid*, pp. 344–345; "Gunpowder", *ibid*, pp. 389–390; "Guo Shoujing", *ibid*, pp. 390–391; "Li Zhi", *ibid*, pp. 513–514; "Liu Hui and the *Jiuzhang suanshu*", *ibid*, pp. 514–515; "Magic Squares in Chinese Mathematics", *ibid*, pp. 528–529; "Navigation in China", *ibid*, pp. 762–765; and "Yang Hui", *ibid*, pp. 1043–1044.

the traditional Chinese concept of the term *shuxue* 数学 that is used as the modern equivalent for "mathematics".

In July I went to Singapore to deliver the 1997 series of Wu Teh Yao 吴 德耀 Lectures.[132] Besides giving a staff seminar at the Department of Chinese Studies of the National University of Singapore, I was a guest at the official opening of the East Asia Institute with Wang Gungwu as its Director. On 20 July Tan Sri Tan Chin Tuan, the biggest donor to the Needham Research Institute to date, invited me for luncheon at his residence together with his family and some mutual old friends. The next evening I delivered a public lecture in Chinese on "Zhuge Liang and the magical art of *qimen dunjia* 奇 门遁甲 " to an audience of some 500 in an auditorium that could sit only 400. In the evening of 21 July I was the guest of honour at a dinner hosted by the Wu Teh Yao Memorial Lectures Foundation. Among the guests were Mr George Hicks and his wife Mrs Judie Hicks, and Ms Lee Siok Tin. I knew the Hicks whilst I was a director of the Hong Kong East Asian History of Science Foundation. The Hicks rendered a helping hand in raising funds for Needham besides contributing money of their own. Lee Siok Tin and I graduated together in 1950 with First Class Honours Bachelor degrees. Her father Dr Lee Kong Chian made a "munificent" donation to the *Science and Civilisation in China* project in the late 1950s, and the Lee Foundation, which is a family Trust founded by Dr Lee, makes annual contributions to the Needham Research Institute. The next day her brother Lee Seng Tee invited me to lunch at the Lee Foundation Luncheon Club, where I met his elder brother Lee Seng Gee and several old friends. That evening I gave a public lecture in English entitled "Did Confucianism hinder the development of science in traditional China?" to a full house in a lecture theatre.

While in Singapore I received a letter from Dr Catherine Jami, President of the History of East Asian Science Society, about the difficulty of finding someone to organise the proposed 9th International Conference of

---

[132]See Ho, P.Y. (1998), "*Zhuge Liang yu qimen dunjia* 诸葛亮与奇门遁甲 " (Zhuge Liang and his magical art of *qimen dunjia*, in Chinese), *1997 Wu Teh Yao Memorial Lectures*, Singapore and Ho, P.Y. (1998), "Did Confucianism hinder the development of science in China?" *1997 Wu Teh Yao Memorial Lectures*, Singapore.

the History of East Asian Science. It so happened that the Dean of the Faculty of Arts and Social Sciences invited me to lunch and I mentioned to him casually about the letter. To my surprise he spontaneously offered to hold the conference in Singapore in 1999 on condition that I sat on its International Advisory Board. As President of the East Asian History of Science, Technology and Medicine Society, Catherine Jami had been trying very hard to get the conference organised in Hong Kong, but without success. Here in Singapore things turned out quite unexpectedly, with someone readily agreeing to hold the conference without any persuasion. This is how things happen sometimes.

On 31 July I was elected academician by the General Assembly of the newly established International Eurasian Academy of Sciences, a non-government UNESCO organisation with regional centres in Tokyo, Beijing, Cairo, Moscow, Paris and Dresden. By this time I was already back home in Brisbane. I normally went from Brisbane to London via Singapore, taking a direct Singapore-to-Heathrow flight. This time I thought of making the trip via Beijing for a change to give me the opportunity to see a little more of China. The trip first took me to Hong Kong on 22 September. There I had something to attend to on behalf of the Needham Research Institute. I met with Sir Q.W. Lee and his team over lunch at the penthouse of the Hang Seng Bank and listened to his views on the negotiations with Totem, the filmmaker, concerning the Dragon's Ascent project. Sir Q.W., as we usually called him, raised US$5 million single-handedly to serve as capital for the project, $1 million of which came from his own family trust. He formed the Chinese Civilisation and Education Trust in Hong Kong to look after the funds. Geoffrey Lloyd and I were among his list of members of trustees. He and his team of financial experts at the Hang Sang Bank supervised and scrutinised all expenditures incurred on behalf of the project. He assured me that his object was to look after the interests of Cambridge and Beijing as well as to be answerable to the donors. Thanks to him the whole project was completed by the year 2000. The film found a buyer the following year and the sum accrued, after deducting expenses, was equally divided between the Needham Research Institute and the Institute for the History of the Natural Sciences in Beijing. The following day I had lunch with Dr Philip Mao and

Dr Peter Lee at the Kowloon Hotel. They discussed with me the future of the Hong Kong Trust and the need to bring in new blood. It was at this juncture that Philip Mao indicated to me his decision to retire from his chairmanship of the Hong Kong East Asian History of Science Foundation. I reported to him my meeting with Sir Q.W. Lee, whom he knew as a friend. The same evening I joined Mr P.L. Lam and Mrs Mary Lam at dinner. They were both old friends of Needham and had contributed substantially to him and his *Science and Civilisation in China* project.

I arrived in Beijing on 26 September and attended a welcome dinner hosted by the newly appointed Director of the Institute for the History of Natural Sciences, Professor Liu Dun, and his two Deputies. The next day I attended a gathering to celebrate the 40th anniversary of the Institute for the History of the Natural Sciences and made a congratulatory speech on behalf of the Needham Research Institute. At the gathering I also met Dr Lu Yongxiang , the new President of the Chinese Academy of Science. Professor Liao Ke 廖克, President of the China Science Centre of the International Eurasian Academy of Sciences, visited me at my hotel to present to me the Diploma of his Academy. On 28 September I went with Dr Hu Weijia 胡 维佳 to Tiananmen to visit the Museum of Historical Relics, where the exhibition of recent archaeological findings was held. The next day I received the 1995 Award from the Commission of Science and Technology for International Scientific and Technological Cooperation on behalf of the late Joseph Needham. The ceremony took place at the VIP Room, Friendship Hotel, Beijing. On 30 September I gave a seminar at the University of Science and Technology, Beijing, to the research group under Professor Ke Jun and Professor Han Rubin 韩汝玢.

On 1 October an official car from the Institute for the History of the Natural Sciences took me along a recently constructed highway to Taiyuan. On arrival Professor Guo Guichun, Executive Vice-President of Shanxi University, and Professor Qiao Shouning, Director of the Foreign Affairs Office of the university, extended to me their welcome to stay at the university guesthouse. The next day Professor Guo took me on a tour to places of interest near the city. On 3 October I left Taiyuan for Xi'an by train, accompanied by Mr Yao Licheng, the Assistant of Professor Liu Dun. The next morning Professor Qu Anjing 曲安京 met us on our arrival at Xi'an and took us directly to the North-West University to meet his research students. That afternoon I delivered a public lecture at the university on the method of *qimen dunjia* practised within the Chinese astronomical bureau for weather forecasting and for divination.[133]

On 5 October I travelled to Luoyang by train in the company of Professor Qu Anjing. My purpose was to visit two old friends from my home State of Perak who went to settle down in China in the early 1960s. They were Dr Wang Qinghui (formerly known as Ong Cheng Hooi 王清辉) and Professor Feng Cuihua 冯翠华 (formerly known as Anita Fung). I worked together with Cheng Hooi in 1949 to raise funds for the University of Malaya when I was Chairman and he the Honorary Treasurer of the Ipoh Graduates and Undergraduates Dance Committee. We succeeded in raising the largest amount of money among similar efforts throughout the country. That was the only time I had contact with him until this visit. Anita was the Arts students' representative in the Raffles College Table-Tennis Committee of which I was chairman and college captain in 1948 and 1949. When I visited them Cheng Hooi had already retired from active service as a doctor and was in poor health suffering from emphysema. Anita was a professor teaching English at the PLA Language College holding the equivalent rank of a Major-General. Their living quarters was a double-storey semi-detached house, somewhat like the university quarters I once stayed in at College Green in Singapore. They had their own bicycles, but Anita could use the service of a military car driven by a young PLA member. Cheng Hooi used to work as an army doctor and rendered help to translate medical articles from Chinese to English to be published overseas. Anita was a popular and much respected teacher. She wrote a book on rhetorical English that sold tens of thousands of copies. As she had to take a class, she let me use her official car with a PLA chauffeur to take me sightseeing.

---

[133] See Ho, P.Y. (1998), "*Cong kexueshi guandian shitan Qimen dunjia*" (An Examination of the traditional Chinese secret magical art of *qimen dunjia* from the standpoint of history of science, in Chinese), *Xibei daxue xuebao*, 28.1:1–4 and 28. 2:93–97.

Cheng Hooi and Anita had to leave Singapore in 1962 for political reasons. According to them, they could return to Malaysia though, if they wished to, because Cheng Hooi's friend and college mate, Dr Mahathir, the then Prime Minister of Malaysia, could make it possible. However, both of them had got used to the life in China and had a son who was an officer in the PLA. Cheng Hooi told me that in retrospect he thought that Lee Kuan Yew had done better for Singapore than his own group of politicians could have done. That was the last time I saw Cheng Hooi. Nine months later I received news from Anita that she had lost her husband. I returned to Beijing by train on 6 October. The next day I took a BA flight for Heathrow.

Much of the energy of the Needham Research Institute at this time went to the Dragon's Ascent project. I once made a comparison between that project and the World Soccer Competition while talking to John Moffett, the Librarian of the Institute, who used to follow the games. The success of both depended very much on teamwork. Getting Sir Q.W. Lee to raise funds for the project was equivalent to getting into the qualifying rounds in soccer. Success in raising the required amount for the project would get one into the quarter-finals. The production of the film and its quality amounted to the semi-finals. Then came the sale of the film, which would be the finals in the match. I said that we had already got past the qualifying rounds and the quarter-finals, and were at that time playing hard in the semi-finals. The filming of the project was the responsibility of Totem, the filmmaker, with the Institute playing an important role in the content of the film as well as liaison between Beijing, Hong Kong and the filmmaker. Here Christopher Cullen should take most of the credit. While Cullen was involved in all the four stages, I played the greatest part in the qualifying rounds, while Geoffrey Lloyd later took over in the finals. All I could do for in the Dragon's Ascent project at this time was just to listen and offer encouraging words. I returned to Brisbane in December.

In April 1998 I went to Taiwan via Hong Kong, where I stopped over to conduct the oral examination as the external examiner for a Ph.D. candidate at the Department of Architecture of the University of Hong Kong. I also gave a talk on the Chinese *jiugong* 九宮 magic square and ancient Chinese architecture. In Taiwan I visited the Tsinghua University in Hsinchu before attending two review board meetings of the Chiang Ching-kuo Foundation. I made a trip to Tokyo and Kyoto before returning to Brisbane via Singapore. On 1 May I left Brisbane again, this time for Cambridge.

The *Nihon Keizai Shimbun* 日本经济新闻, Inc. and the Centre of Cutting-Edge Research in Science and Technology, University of Tokyo, jointly organised an "Asian Science and Technology Conference" and invited Dr Yuan Tseh Lee 李远哲, Nobel Laureate and President of Academia Sinica, Taipei, to be a keynote speaker. Dr Lee wrote to ask me to take his place. Accordingly I received an invitation from the organisers to present a paper and to speak in a panel. I left London on the last day of May for Tokyo, arriving on 1 June. That evening my host gave me a welcome dinner and invited my two friends Shigeru Nakayama and Teruyo Ushiyama as guests. On 2 June I gave my keynote speech and participated in a panel discussion. I returned to Cambridge immediately after the conference. The Jiaotong University, Shanghai, appointed me Honorary Professorial Advisor for its newly established Department of History of Science and Philosophy.

In July 1998 Lucy and I went to Taipei, where I attended the Academia Sinica meeting at Nankang. We visited San Francisco to see Lucy's relatives before returning to Brisbane. I returned to Cambridge on 29 August. On 27 to 30 October I attended the 3rd International Conference on the History of Oriental Astronomy at Fukuoka.[134] I stopped by in Hong Kong to see Sir Q.W. Lee about the progress of the Dragon's Ascent project and to talk with the Hong Kong Trust about the arrangements for me to deliver the Needham Memorial Lecture in 1999. I then returned to Brisbane.

In March 1999 on my way to Taipei I stopped over in Singapore to advise the organisers of the 9th International Conference on the History of Science in East Asia to be held in Singapore later in the year. On arrival in Taipei on 25 March I gave the plenary lecture at the 5th Symposium of History of Chinese Science at the Academia Sinica and served as discussant

---

[134] See Ho, P.Y. (2000), "Western influence on indigenous Chinese astrology", Masanori Hirai, ed., *Proceeding of the Third International Conference on Oriental Astronomy, Fukuoka, 1998*, Fukuoka, pp. 221–225.

for a panel of presentations on divinatory arts. The meetings of the Chiang Ching-kuo Foundation review boards took place over the weekend on 3 and 4 April. The next day I left for Tokyo and Kyoto to consult Shigeru Nakayama and Michio Yano on some points concerning the book *Chinese Mathematical Astrology: Reaching out to the Stars* that I was then writing. In Kyoto Yano took me to visit Kiyosi Yabuuti at his home. Yabuuti presented me with two new editions of his books. That was the last time I saw this great Japanese historian of science. From Japan I went to Malaysia and visited Penang on 28 April. I was the house guest of the Head of State, Hamdan, at Seri Mutiara. After attending the wedding ceremony of my nephew Yik Tuck 奕 德, I left for London via Singapore.

On my return to Cambridge on 3 May, I found both Cambridge University and the Needham Research Institute preparing for the visit of President Jiang Zemin from China. I met President Jiang at the Cambridge University Library, where the Needham Research Institute presented him with a set of the *Science and Civilisation in China* that had been published so far. I was also a guest at the lunch given in honour of President Jiang at Madingley Hall by the Vice-Chancellor of Cambridge University.

The 9th International Conference on the History of Science in East Asia took place at the Merchant Court Hotel in Singapore on 23–27 August 1999. I spoke at the opening ceremony and requested the Tan Kah Kee Foundation to host a welcome dinner for the delegates from China. It was then decided that the 9th International Conference of the History of East Asian Science would be held in Shanghai some time in July 2002.

In December 1999 I delivered the Second Joseph Needham Memorial Lecture for the Hong Kong East Asian History of Science Foundation at the Furama Hotel, Hong Kong. I spoke on Hong Kong Radio 2 and participated in the production of the Hong Kong government TVE education programme for teachers of secondary schools. I also met Mr Victor L.L. Chu 诸立力, who invited me to lunch at the Hongkong Club. He suggested that the Needham Research Institute should participate actively at the 9th International Conference of the History of East Asian Science in Shanghai and promised to contribute HK$50,000 to assist the Institute for that purpose.

In 2000 I went to Hong Kong to attend a conference on Chinese studies in the 21st century held between 17 and 19 January at the University of Hong Kong. As the Hong Kong East Asia History of Science Foundation wished to arrange for a meeting between me and David Li, a luncheon meeting was scheduled for Friday, 12 January at the Bank of East Asia. It turned out that a memorial service at St. John Cathedral for the late Lady Ivy Fung had also been arranged for 12 noon that same day. The Fung family had been strong supporters of the University of Hong Kong, contributing the Fung Ping Shan Library, the Fung Ping Shan Museum and The Fung Ping Shan Scholarship. Fung Ping Shan was the founder of the Bank of East Asia, of which Dr David Li's grandfather was one of the three partners. Lady Ivy Fung was the daughter-in-law of Fung Ping Shan and the late wife of Sir Kenneth Fung Ping Fan 冯秉芬, Lucy's fourth uncle. I went with the Acting Vice-Chancellor of the University of Hong Kong and the Acting Head of the Department of Chinese of that university to attend the service. David Li was also at the same memorial service and our meeting did not start until we got back to the bank at 1.15 pm. There was only one unwritten item on the agenda concerning the NRI, while I answered questions after reporting on the latest about the Dragon's Ascent according to what Sir Geoffrey Lloyd had updated me the day before over the telephone. Much interest, particularly from Peter Lee was expressed in the appointment of the new Director. Dr David Li said that he would like to speak personally to Sir Geoffrey Lloyd during his coming visit to Cambridge on 17 March. I reminded him that the final decision belonged to the UK Trustees.

In April 2000 I went to Taipei for the meetings of the Chiang Ching-kuo Foundation review boards. Professor Chu Yun-han 朱云汉 took over the post of Executive-President when Li Yih-yuan became Chairman of Directors of the Foundation. After visiting the National Tsinghua University at Hsinchu, I made a trip to Tokyo and Hong Kong before returning to Brisbane via Singapore.

After a short stay in Brisbane I went to Cambridge. Lucy joined me later in May. In late June Lucy and I travelled to Taipei via San Francisco, where we had a short stay with Lucy's relatives. After the meeting of Academia Sinica we went to Hong Kong, where I met some members of the Hong

Kong Trust. We then left for Singapore. Lucy stayed behind to attend a niece's wedding ceremony, while I took off for Cambridge on 11 July. The filming of the Dragon's Ascent was completed. Sir Q.W. Lee and Geoffrey Lloyd were trying hard to find a buyer. I did not get myself involved at this difficult final stage. That set me free to write a book before I relinquished my post as Director of the Institute. The manuscript for my book was submitted to the editor, Christopher Cullen, for inclusion in the *Needham Research Institute Monograph* series.

I visited Kaohsiung, the second largest city in Taiwan, to participate in the opening ceremony of the exhibition on "Dr. Joseph Needham in War-time China (1942–1946)" on 7 December 2000 to mark the 100th birthday of the Founder of our Institute. I read a paper at a symposium on Needham that took place in conjunction with the exhibition.[135] Professor Wu Chia-li 吴嘉丽, a professor of chemistry who pioneered teaching of the history of science in Taiwan and whose husband was a senior and influential government official in Kaohsiung, expressed the view that the Needham Research Institute should be linked with Cambridge University in order to gain financial support from Taiwan. I replied that we were all working towards that direction. I gave a lecture to a class of some 400 undergraduates at the Sun Yat-sen University and was introduced to the audience by Professor Chung Ning 锺玲, the Dean of Arts.[136] After saying that I was a member of Academia Sinica and the Director of the Needham Research Institute, she asked those who had heard of Academia Sinica to raise their hands; I noted that all hands were raised. Then she asked again for those who had heard of Needham's name to raise their hands, but not a single hand was raised. It gave me the opportunity to invite my audience to visit the exhibition in the Museum in order to know more about Needham. This little incident

---

[135] See Ho, P.Y. (2000), "My collaboration with Dr Joseph Needham on the History of Chinese Science", *Proceedings of the Symposium on the History of Chinese Science to mark the 100th birthday of Dr Joseph Needham, Kaohsiung, December 2000.*

[136] She was a former colleague of mine at the Department of Chinese, University of Hong Kong.

indicated how timely the exhibition was and how much more there was to be done to publicise Needham's name.

My last year of service at the Needham Research Institute was a busy one. The Liaoning Press published my selected Chinese works.[137] On 17–19 January 2001 I attended the "International Conference on Chinese Studies in the 21st Century: A New Vision" at the University of Hong Kong, and delivered a paper entitled "The Reunion of Science and the Humanities in the History of Chinese Science". I returned to Cambridge on 19 February 2001. In the same month I participated in the 2001 Wolfson College Lecture series at Oxford University, on China's technology transfer to the world. I delivered a lecture, entitled "The Gunpowder Epic". Rayson Huang came all the way from Birmingham as a guest of Wolfson College to attend my talk. I next attended an International Conference on Chinese Studies at the University of Hong Kong. Towards the end of March 2001 I made a trip to Taiwan to be the keynote speaker at the International Symposium on the Chinese History of Science at Tamkang University, Tanshui, and to give a lecture to the students of that university as well.[138] I then attended two review board meetings of the Chiang Ching-kuo Foundation to make recommendations for the granting of financial support to universities and scholars in Europe and the Pacific Rim.

In June I went to Malaysia on vacation and took the opportunity to catch up with potential donors to Cambridge and to give a talk at the Perak Academy in Ipoh. In September I was the keynote speaker at the International Conference on Zheng He 郑和, the early 15th-century Chinese navigator and took the opportunity to discuss about likely mutual cooperation with Cambridge to hold the next conference on Zheng He. The National Tsinghua

---

[137]See Ho, P.Y. (2001), *Science and Culture in Ancient China: Selected Works (Written in the Chinese Language) of Ho Peng Yoke,* Liaoning.
[138]In Ho, P.Y. (2001), "*Kexuejia de zhiyi jingshen: Shen Gua yu Liuren shushu* 科学家的质疑精神: 沈括与六壬术数" (The spirit of scepticism of a scientist: Shen Gua and the magical art of *Liuren,* in Chinese), *Proceedings of the International Symposium on the Chinese History of Science, March 24–25, 2001,* Tamkang University, Taipei, pp. 5–12.

University at Hsinchu offered me an appointment as endowed Honorary Visiting Professor for a period of three years from September 2001 to August 2004. In October I attended the 9th International Conference on Chinese Science held at the City University of Hong Kong. That was also the last time that I met members of the Hong Kong East Asian History of Science Foundation officially. In the same month Professor Chang Chi-kang 张志刚, President of the City University of Hong Kong, visited the Needham Research Institute with a team of four that included Professor Cheng Pei-kai 郑培凯, Director of the Centre of Chinese Culture, of the same university. They requested me to give a series of lectures on behalf of the Needham Institute at their university in Hong Kong. I agreed to do so in memory of my late friend Dr Cheng Te-k'un, who was together with me in Cambridge, Kuala Lumpur and Hong Kong, and was a scholar much respected by Joseph Needham. I presented my last Director's Report to the East Asian History of Science trustees' meeting that was held on the last Friday afternoon of October. My directorship ended on the day of that trustees' meeting, although I remained a trustee for another year.[139] The last section of my report to the Trustees said:

> Dr Needham had made sure that I would not be bogged down by his *Science and Civilisation in China* project like he was, by approaching someone else to be the coordinator of the project before I came to take over the directorship from him. I am glad that the project is now in the capable hands of Dr Christopher Cullen, leaving me free to pursue what I think is good for the Institute in consultation with the Chairman of the Trustees and my small team of NRI colleagues. I wish to thank all members of the UK Trust for the amicable manner we have worked together during the past 12 years. Although my appointment is non-stipendiary, I consider working with a mutually understanding team towards a common goal in the congenial Cambridge surroundings in itself sufficiently rewarding.

---

[139] I was later reappointed as a Trustee until 2007.

Before I left Cambridge I was conferred Emeritus Director of the Needham Research Institute, and a workshop on the history of divination was held in my honour.

After Cambridge

The conclusion of my career as Director of the Needham Research Institute was soon followed by the ending of a dream that I would be able to continue playing some role to serve the education of future scholars on a wider international scale. On 23 May 2000 Wu Chang-sheng (Fred) visited me at the Needham Research Institute together with his two sisters Yu-lin and Ellen (Yu-chen 玉珍), and another friend. Fred studied law in Cambridge in the late 1950s at Emmanuel, the same College that his father Wu Lien-teh attended more than 50 years ago. He became a very successful barrister in Singapore and contributed generously to the building fund of Emmanuel College. In his profession he was known as C.S. Wu. In his younger days he accompanied his father to seek funds from friends to help the *Science and Civilisation in China* project of Joseph Needham. During our conversation concerning how the Needham Research Institute helped scholars from China, Fred was impressed to hear how we managed to get around the red tape in China and related what happened recently to a client of his. That client was a Chinese lady from Indonesia, who had sought his legal advice to establish a trust worth billions of dollars for the purpose of providing opportunities to scholars in China to go abroad for higher education. She would like Fred to be a trustee. Fred said he would agree provided that Chinese bureaucracy was not involved. She approached the Chinese authorities and became disillusioned, so she did not proceed with the formation of the trust. I suggested that our Cambridge experience might provide a way out. Fred said that he would arrange for me to meet her during one of my future visits to Singapore. If successful I would have a new role to play in the promotion of higher education after my retirement from Cambridge.

The next year I wrote to Fred about my intended visit to Singapore. He replied that we should meet this lady soon because of her advanced age. In August I saw Fred in Singapore. He invited me to dinner and to join him to travel to Bangkok by train on the *Orient-Express* as his guest. He seemed to have forgotten about meeting the old lady. I was unable to join him on his trip to Bangkok. He went in September and fell down during the journey. On his return to Singapore he was found to have an advanced brain tumour and was admitted to the Mount Elizabeth Hospital. I visited him there in October and found him in a coma. That was the last I saw of Fred; he passed away in February 2002 and so ended my dream. True to his profession, Fred did not divulge the identity and address of his client.

This was another case of "man proposes, God disposes", with the Chinese equivalent "*moushi zai ren chengshi zai tian*", although it had no direct relation with Needham. I just took it as a natural cause of events. Considering my age this might turn out to be to my advantage. I therefore decided to spend my retirement writing. The first book I wrote was *Chinese Mathematical Astrology: Reaching out to the Stars*, published in the *Needham Research Institute Monograph* series by Curzon/Routledge in 2003. The second is *Explorations in Daoism: Science in Literature*, to be jointly published in the *Needham Research Institute* and the *Studies in Buddhism* series (forthcoming). The present book is the third to be submitted for publication, but will probably be off the press earlier.

On 8–12 April 2002 I was in Beijing attending the International Symposium on Retrospective and Perspective of Science in Modern and Pre-Modern China at the invitation of Professor Song Jian 宋健, President of the Chinese Academy of Engineering. I spoke on the forthcoming concluding volume of *Science and Civilisation in China*. On 12–13 April I attended my last meetings of the Chiang Ching-kuo Foundation review boards after a service of 12 years. I visited the National Tsinghua University at Hsinchu from 14 to 16 April in my capacity as endowed Honorary Professor.

In July I attended the 26th meeting of Academia Sinica members in Nankang. I stopped over in Japan to find out the result of my attempt with some Japanese scholars to raise funds to establish a Yabuuti Lecture at Cambridge University, and to consult Shigeru Nakayama on the ownership of an early 19th-century transcribed Chinese alchemical text. I then stopped over in Hong Kong on my way back to Brisbane, where I delivered a series of three lectures at the City University of Hong Kong.[140] It was the first time I gave a public lecture in Cantonese in Hong Kong. I found that talking to people about their own culture in their mother tongue or in a foreign language made a difference. My host, Cheng Pei-kai 郑培凯 vividly described the large gathering of my audience in his edited book *New Perspectives on the History of Science in China: Divination, Astronomy, and Medical Science* (Hong Kong). Between 14 and 22 October I was in Beijing to attend the annual meeting of the Beijing Centre of the Eurasian Academy of the Sciences and to visit the Institute for the History of the Natural Sciences. At the latter I gave a talk to introduce a new book that I was writing, and discussed with Liu Dun fundraising matters concerning the 2005 International Congress of History of Science that would be held in Beijing. Towards the end of October I attended the Trustees' meeting at the Needham Research Institute. I was a guest of David McMullen staying at St. John's College, Cambridge.

In April 2003 Lucy and I visited Singapore and Kuala Lumpur. From Kuala Lumpur we went by car to Ipoh, where I gave a talk at the Perak Academy. I also gave a lecture at the new Universiti Tunku Abdul Rahman in Kuala Lumpur. Unfortunately the news of SARS was floating around in Malaysia and Singapore, preventing us from seeing some friends, but this was more than compensated by our reunion with Ho Chuen and his wife Cecilia, who came all the way from Seremban to meet us. In July I went to the National Tsinghua University at Hsinchu to perform my duty as endowed Honorary Professor. My duties involved meeting graduate students, listening to their presentation of work-in-progress reports and giving

---

[140] See "*Zhongxi wenhua jiaoliu zhong de jiugongtu yu mofangzhen* 中西文化交流中的九宫图与魔方阵 " (The *jiugongtu* and magic squares in the cultural transmission between China and the West), You wenhua jiaoliu chansheng de *ziwei doushu* 有文化交流产生的紫微斗数 (The *ziwei doushu* astrology as a product of intercultural exchange), and "*Tan Sanguo yanyi* zhong de Zhuge Liang yu *qimen dunjia* 谈三国演义中的诸葛亮与奇门遁甲 " (On Zhuge Liang and the *qimen dunjia* magical art as described in the *Romance of the Three Kingdoms* novel), in P.K. Cheng, ed. (2003), *New Perspectives on the History of Science in China: Divination, Astronomy, and Medical Science*, (in Chinese), City University of Hong Kong Press, Hong Kong, pp. 3–16, 17–29, 30–44.

comments. I then went to Japan to try and get some news about the effort to raise funds for the Kiyosi Yabuuti Lecture at Cambridge. Nakayama and his wife came to see me at my hotel.

My book *Chinese Mathematical Astrology: Reaching out to the Stars* in *Needham Institute Monograph Series* by Curzon/Routledge was published in 2003. In October I travelled by Singapore airlines to Heathrow and heard that the Airlines would be bringing Mrs Lee Kuan Yew back to Singapore from London because of illness. I left a note with the airlines wishing her a speedy recovery. She was my senior contemporary at Raffles College in 1946. Her husband acknowledged my note on her behalf. (See Figure 23.) I stayed at a Fellow's Guest Room at Robinson College. I also attended a College dinner at Caius College, sitting next to Professor Stephen Hawkins. As he was occupying Room K1, which was formerly used by Needham, our conversa-

Senior Minister
Singapore

20 November 2003

Dear Ho Peng-Yoke

Thank you for your letter of 29 October and your good wishes for my wife.

Yours sincerely

Professor Ho Peng-Yoke
Emeritus Director
8 Holdway Street
Kenmore, Queensland 4069
Australia

Figure 23 Letter from Lee Kuan Yew

tion naturally turned to Needham. Talking through his speech machine, he mentioned the four Chinese characters "*ren qu liu ying* 人去留影" that Needham wrote on the wall, indicating that although he had left, his influence still remained. After attending the Trustees' meeting at the Needham Research Institute and submitting the manuscript for my new book *Explorations in Daoism: Science in Literature* to the editor, Christopher Cullen, for publication, I returned to Brisbane via Singapore.

In April 2004 I went to Taiwan to complete my last term of duty as endowed Honorary Professor at the National Tsinghua University, Hsinchu. I then went to Tokyo to visit Shigeru Nakayama and his family. They took me to dinner at a restaurant in Ginza. Nakayama and I were amazed to find that parts of Ginza had changed beyond recognition since the last time we were there. I also had dinner with Teruyo Ushiyama. Both Nakayama and Ushiyama were old friends of Needham, whom we talked about whenever we met together, dubbing our meeting a mini Needham symposium. On my return from Tokyo I stopped over in Singapore and Kuala Lumpur to visit some relatives and friends. In Kuala Lumpur I gave a public lecture on the link between Islamic studies and Chinese studies that was organised jointly by the Centre of Malaysian Chinese Studies and the Soka Gakkai in Kuala Lumpur. I stressed that inter-cultural studies were not only important in inter-disciplinary research, but also for the promotion of mutual understanding in a multi-cultural world.

After attending a meeting of the Academia Sinica in Nankang in July, I went straight to Singapore. My purpose of coming to Singapore then was to fulfil a promise I made to Dr Phua Kok Khoo 潘国驹, the Editor-in-Chief and Chairman of World Scientific Publishing Co. to give a talk about myself in Singapore. He had earlier asked me to write a book about myself for publication, but I could not decide whether the book should be written in English or Chinese, or in two different versions, one in each language. Phua and I had a mutual friend in the Nobel Laureate Professor C.N. Yang 杨振宁. During an interval at the Academia Sinica meeting in Nankang I had a chat with C.N. Yang, talking about Phua and our coming meeting in Beijing at the International Congress of History of Science in 2005. Yang asked me to convey to Phua his felicitations when I met him in Singapore. As an

Adjunct Professor at the Physics Department of the National University of Singapore, Phua made arrangements with Professor Oh Choo Hiap 胡祖协, the Head of the Physics Department, for me to give a talk at the Faculty of Science auditorium. They suggested that the talk be entitled "From Physics to Chinese Studies: What can a Physicist do in Chinese Studies that a Sinologist is unable to do?"

At my talk I stressed on the word "adaptability" and reminded my audience what their predecessor, Professor J.C. Cooke said half-a-century ago at a meeting of the Faculty of Science, that the Faculty was not meant to train students in any particular trade, but to enable them to fit into all things in life. After roving in four continents, I found myself returning to Singapore to give a talk involving physics, the humanities and Joseph Needham at the very same Physics Department where I first began my academic career. When I worked in Singapore and served as Chairman of the Faculty of Science Public Lectures Committee, I used to get others to speak, but half-a-century later I had to play the role of the speaker. Things seemed to have come full circle for me. This is an opportune moment to conclude this book with the hope that it would make unusual reading as was the title of the talk I gave in Singapore.

Index

C

Printed in the United States
By Bookmasters